CELLULAR AUTHENTICATION
AUTHENTICATION
FOR MOBILE AND INTERNET SERVICES

CELLULAR AUTHENTICATION

FOR MOBILE AND INTERNET SERVICES

Silke Holtmanns, Valtteri Niemi, Philip Ginzboorg, Pekka Laitinen and N. Asokan

All at Nokia Research Center, Helsinki, Finland

A John Wiley & Sons, Ltd, Publication

Library of Congress Cataloging-in-Publication Data

Cellular authentication for mobile and Internet services / Silke Holtmanns . . . [et al.]
    p.   cm.
  Includes bibliographical references and index.
  ISBN 978-0-470-72317-3 (cloth)
  1. Mobile communication system–Access control.   2. Internet–Access control.
3. Cellular telephone systems–Access control.   4. Wireless LANs–Access control.
5. Authentication.   I. Holtmanns, Silke.
  TK6570.M6.C44 2008
  005.8 – dc22

                                                                      2008018998

A catalogue record for this book is available from the British Library.

ISBN   978-0-470-72317-3 (HB)

Set in 11/13pt Times by SNP Best-set Typesetter Ltd., Hong Kong
Printed in Singapore by Markono Print Media Pte Ltd

Contents

Preface

Like many useful and successful systems, the Internet was not designed with security in mind. Consequently, authentication on the Internet has been a vexing problem ever since it was opened up to the larger public in the early 1990s. The fact that most computers on the Internet are general-purpose devices has made this problem more acute. Peter Steiner's famous New Yorker cartoon with the caption 'On the Internet, nobody knows you are a dog' continues to remain apt. However, there is an existing useful and successful system that has struck the right balance among security, cost, and usability: this is the worldwide cellular authentication infrastructure. This is the infrastructure that allows a mobile network subscriber to travel halfway across the world, have the local mobile network operator authenticate the subscriber, provide cellular services, and subsequently charge him for those services. Naturally, many people have wondered whether it is possible to use this large-scale infrastructure to secure services on the Internet. Already in the late 1990s, there were examples of using short message service (SMS) messages to pay for soda on a vending machine. These are ad hoc means of bootstrapping security for new applications from the existing cellular security infrastructure.

Back in 2000, we started to think about designing a systematic approach to bootstrap cellular authentication for Mobile and Internet Services. What began as a small Nokia Research Center project grew into a standardization work item in 3rd Generation Partnership Project (3GPP). The first version of the standardization of the basic Generic Authentication Architecture (GAA) features was completed in 2006. Two lead applications for GAA emerged: Multimedia Broadcast/Multicast Services specified by 3GPP and Smart Card Profile for broadcast Mobile TV, specified by Open Mobile Alliance.

There are a number of detailed specification documents of GAA and its applications. In 2005, we published a paper titled 'Extending cellular authentication as a service' at the First IEE International Conference on Commercialising Technology and Innovation. While GAA was being specified, it was of primary interest to engineers involved in standardization. The existing standards documents were sufficient to meet the needs of this group.

In the last two years, several new groups of people have become interested in GAA, including technical architects, software developers, executive decision makers, and academic researchers. So far, there has been no single source for a comprehensive, yet readable treatment of GAA, which will present the technology in a form that is accessible to these groups of people. Our desire in writing this book is to address this gap.

Deployment of broadcast Mobile TV is expected to begin in the latter half of 2008. We expect this will introduce GAA to more people who would want to learn more about the technology. We hope that this book will help them.

Many people helped along the way to take GAA where it is today. Several colleagues at Nokia Research Center contributed to the NRC GAIN project where we conceived and did much of the early work. Standardization delegates from various companies took active roles in developing GAA and its applications, not only in 3GPP SA3 working group, but also in other fora. Several people in Nokia and other companies were instrumental in transforming this technology from research to product. Our sincerest thanks to all of them; without their hard work, we would not have had the possibility to write this book.

Acknowledgements

3GPP™ TSs and TRs are the property of ARIB, ATIS, ETSI, CCSA, TTA and TTC, who jointly own the copyright in them. They are subject to further modifications and are therefore provided 'as is' for information purposes only. Further use is strictly prohibited.

All images of Nokia devices are reproduced with the kind permission of Nokia.

1

Introduction

Over the last few decades, information technology has changed the world in two major ways. The first development is computerization. Computers have gradually entered into almost every walk of life. Many services that used to require interaction with humans are now provided by computers. When you withdraw money from your bank account, or look for a book in the library, or check in at the airport, chances are that you are interacting with computer systems hosting these services.

The second development is the immense and increasing popularity of an open network of interconnecting computers, the Internet. Access to computerized and automated services now takes place over this open network: when you use online banking to make payments from your bank account, or order books from an online bookstore, or browse the online photo albums of your friends, you are using the Internet, as well as a variety of communication links to access the Internet.

In this chapter, we will discuss what is needed to secure the access to applications and services over open networks, what the primary difficulty in achieving security is, and why Generic Authentication Architecture (GAA) helps solve this problem. We will also outline how this book is structured and how to go about reading it.

1.1 Authenticated Key Agreement

Although the nature of accessing services has changed, many of these services require some form of controlled access: for example, only you should be able

Cellular Authentication for Mobile and Internet Services
Silke Holtmanns, Valtteri Niemi, Philip Ginzboorg, Pekka Laitinen and N. Asokan
© 2008 John Wiley & Sons, Ltd

to make payments from your bank account. Usually, controlling access to services is contingent on identifying who is requesting access and verifying the requestor's identity. In other words, the serving computer has to *authenticate* the requestor.

In a closed network, like the plain old telephone system, authentication can be implicit based on the presumed physical security of the network. But in an open network like the Internet, physical security is not relevant – it is easy to claim any identity towards a distant server. Worse still, it is easy to pretend to be a distant server towards an unsuspecting client (for example, using IP address spoofing). Therefore, we need to make use of cryptographic techniques for *mutual authentication* in order to have sufficient trust in the authentication process.

In open networks, authenticating the parties at the beginning of a communication session is not sufficient: An attacker may wait for authentication to complete and then hijack the session by inserting, modifying or deleting the messages being exchanged. To prevent this, the authentication process should also establish *session keys* which can be used to guarantee the integrity of the entire communication session.

In some services, the messages exchanged may need to be private. For example, suppose you have an online photo album accessible only to family and friends. When a friend is legitimately viewing the pictures on your album, the information has to travel from the album server to your friend's browser. It may traverse several communication links with varying levels of physical security. For example, your friend's computer may be connected to her access router over an open wireless link. You do not want anyone eavesdropping along the way to be able to see your pictures. Cryptographic techniques for encryption can protect the messages while en route. Session keys established during the initial authentication can be used for encrypting messages exchanged during the session.

The process of mutual authentication and session key agreement is known as *authenticated key agreement*.

1.2 The Challenge in Authenticated Key Agreement

Mechanisms and protocols for performing authenticated key agreement are well known. For example, every time you access a protected web server, your browser and the web server engage in the Transport Layer Security (TLS) handshake protocol for authenticating the server and agreeing on a session key. The challenge in the authentication is in the task of initializing the necessary credentials at the parties involved in the authentication.

Consider what typically happens when you enrol into the authentication system in the bank. You have to visit the bank in person, and possibly show some photo identification to open an account and provide a mailing address. The bank will then send you the credentials needed to access the bank account, for example, a bank card and the personal identification number (PIN) in separate mails. The process costs time and money. One approach to reduce the cost of initialization is to relax the expected level of security and usability. This is the approach taken by popular free e-mail services: initialization is done by the user visiting a signup page and choosing a username and setting a password. The user ends up using the same password for many different services or puts up with the inconvenience of remembering many different passwords. This may in turn cause that users are more frequently calling the help desk, or a special password recovery tool is required, which also introduces costs to the service provider.

An alternative approach is to bootstrap the needed credentials from an existing security infrastructure. One such security infrastructure is the cellular security infrastructure. The cellular security infrastructure has several characteristics that make it a particularly attractive infrastructure to bootstrap security for applications from.

The first and foremost characteristic is its scalability. The Global System for Mobile Communications (GSM) and Universal Mobile Telecommunications System (UMTS) infrastructure consists of hundreds of participating mobile operators and over two billion subscribers worldwide. Most mobile operators have roaming and billing agreements with many other operators. Once you enrol as a GSM / UMTS subscriber with a local operator in your home country, you will be able to authenticate to many mobile operators, and use their networks to make phone calls or send and receive messages.

The second characteristic is its ease-of-use. Cellular authentication is an example of security that remains under the hood and just works. Users are not required to perform any verification or understand technical security concepts.

The third characteristic is its level of security. Authentication in cellular systems is based on the possession of smart cards. Even though some of the cryptographic algorithms in earlier versions of GSM have been broken, the entire system has stood the test of time. GSM / UMTS security architecture is beginning to be acknowledged as an example of the principle of 'good enough' security of striking the right balance among cost-effectiveness, security and usability [Sandhu03].

GAA consists of a set of specifications that describe how the cellular security infrastructure can be used to provide a general-purpose authentication service for applications and services. It has been standardized both in the 3rd Generation Partnership Project (3GPP) and its North American counterpart the 3rd Generation Partnership Project 2 (3GPP2). Deployment of GAA in mobile devices and mobile networks is expected to start in 2008.

The GAA concept is illustrated in Figure 1.1. GAA is a generic architecture for mutual authentication and key agreement (AKA). Its fundamental building block is Generic Bootstrapping Architecture (GBA). GBA enables automatic provisioning of

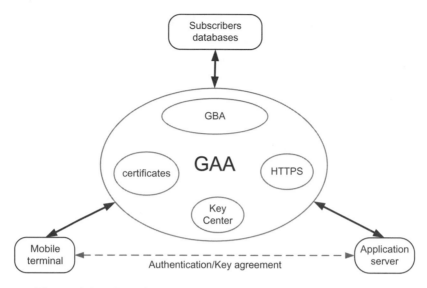

Figure 1.1. Generic Authentication Architecture (GAA) concept

shared keys between the mobile terminal and an application server (provided that the user has a valid subscription[1] to cellular network services).

Other GAA constituents are built on top of GBA (new GAA building blocks continue to be specified):

- Support for subscriber **certificates** (SSC) specifies procedures for the registration of user's public keys. Those procedures are authenticated with GBA.
- Access to application servers with **HTTPS** specifies how to use shared keys created with GBA in conjunction with server-authenticated HTTPS to establish secure and mutually authenticated HTTP communication between mobile terminal and application server.
- **Key Centre** enables creation of keys shared between terminals.

Broadcast mobile television and Multimedia Broadcast/Multicast Service (MBMS), in which encrypted content is wirelessly broadcast, or multicast, are the lead applications driving the deployment of GAA. In those applications, the delivery of service keys that are needed to decrypt the received content is secured with GAA. This makes valid cellular subscription a prerequisite for, e.g., watching mobile television programs. Also, when generating charges for mobile television and MBMS, the person's cellular identity, which is verified with GAA, is used.

---

[1] Valid subscription implies that the mobile terminal and subscriber's databases have a copy of shared key that is used in cellular authentication.

1.3 How to Read this Book?

Our goal in writing this book is to explain what GAA is and how it can be used. We have four different types of readers in mind:

- **Developers** are software designers who design and implement new applications and services. We show how developers can make their application software use GAA for authentication.
- **Architects** are technical experts who design protocols and systems. We explain the GAA concepts and technical details and show examples of how GAA is integrated into existing protocols so that architects can determine if and how GAA could be used to solve the authentication needs of the systems they are designing.
- **Executives** are decision makers in companies who need to figure out whether need to deploy GAA. We provide a general overview of GAA and brief analyses of its benefits and tradeoffs that can serve as background materials for decision making.
- **Academics** are university professors, researchers and students studying computer science or communication systems. We explain the principles and technical details in the design of GAA that can serve as starting points for academics interested in analyzing and evaluating GAA, comparing it to other authentication systems, and designing authentication systems for the future.

The chapters are arranged in the logical order in which we recommend the reader to proceed. Table 1.1 indicates which sections are likely to be of interest to different types of readers.

Table 1.1. How to read this book

Section	Developer	Architect	Executive	Academic
1. Introduction				
1.1 Authenticated Key Agreement	✓	✓	✓	✓
1.2 The Challenge in Authenticated Key Agreement	✓	✓	✓	✓
1.3 How to Read this Book?	✓	✓	✓	✓
2. Classical Approaches				
2.1 Existing Mobile Security Solutions		✓	✓	✓
2.2 General-Purpose Approaches		✓		✓
2.3 Requirements for GAA	✓	✓	✓	✓

Table 1.1. (*continued*)

Section	Developer	Architect	Executive	Academic
3. Generic Authentication Architecture (GAA)				
3.1 Overview of GAA	✓	✓	✓	✓
3.2 Foundations of GAA – Generic Bootstrapping Architecture (GBA)	✓	✓		✓
3.3 Variations of GBA		✓		✓
3.4 Building Blocks of GAA		✓		✓
3.5 Other Architectural Issues		✓		✓
3.6 Overview of 3GPP GAA Specifications	✓	✓		
4. Applications Using GAA				
4.1 Standardized Usage Scenarios (incl. Broadcast Mobile TV)	✓	✓	✓	
4.2 Additional Usage Scenarios	✓	✓	✓	
5. Guidance for Deploying GAA				
5.1 Integration with Application Servers	✓	✓		
5.2 Integration with OS Security	✓	✓		
5.3 Integration with ID Management Systems	✓	✓	✓	✓
5.4 Integration of GAA into Mobile Networks		✓		
6. Future Trends	✓	✓	✓	✓
Terminology and Abbreviations	✓	✓	✓	✓

Reference

[Sandhu03] Ravi S. Sandhu: *Good-Enough Security: Toward a Pragmatic Business-Driven Discipline*. IEEE Internet Computing 7(1): 66–68 (2003).

2

Classical Approaches to Authentication and Key Agreement

In this chapter, we look at existing approaches to AKA in order to set the stage for discussing the motivation for and the design of GAA. We begin by examining the security architectures in mobile networks and then take a broader look at other more general-purpose AKA mechanisms. We then identify requirements for bootstrapping an authentication architecture from existing infrastructures.

2.1 Existing Mobile Security Solutions

In the last two decades, several different mobile network architectures have been designed and deployed. By far, the most widespread have been the UMTS and especially its predecessor GSM infrastructures. Therefore, in this section, we focus on the security of these infrastructures.

2.1.1 UMTS Security Infrastructure

As we pointed out in Chapter 1, there are many compelling reasons for bootstrapping credentials from the existing, widely used, cellular security infrastructures. Before we go on to describe the design and details of the GAA, let us briefly describe the UMTS security infrastructure and the associated terminology in order to motivate the need and the requirements for GAA.

Cellular Authentication for Mobile and Internet Services
Silke Holtmanns, Valtteri Niemi, Philip Ginzboorg, Pekka Laitinen and N. Asokan
© 2008 John Wiley & Sons, Ltd

One of the cornerstones of GAA is the HTTP Digest authentication using AKA protocol (HTTP Digest AKA) [RFC3310]. The UMTS AKA protocol (AKA) has been originally designed for the purpose of securing subscriber access to cellular networks, more specifically, Universal Mobile Telecommunication System (UMTS) or 3G networks [Niemi07], [TS33.102]. The authentication part of the AKA protocol of the UMTS security is needed to verify the subscriber's identity while key agreement is used for generating keys that are subsequently used in encryption of traffic in the radio network and also for protecting integrity of the signalling messages.

These security features are applied for all user data. Therefore, all applications and services used over the cellular access automatically gain a certain basic security level. Because the security protocols and cryptographic algorithms used in UMTS networks are up-to-date with the state of the art in the area, this basic security level is fairly good in many respects. For example, 128-bit keys are used in the cryptographic algorithms and all details of these algorithms are publicly known, and therefore, open for public domain scrutiny as well.

In the rest of this subsection, we give a brief description of the most essential security features in UMTS infrastructure. The UMTS infrastructure consists of a number of mobile operators with UMTS networks. For each subscriber, there is a home network run by the operator with whom the subscriber entered into a business relationship, for example, by opening a billing relationship or purchasing a prepaid subscription. Each subscriber has a unique identifier known as the International Mobile Subscriber Identity (IMSI).

Cutting a few corners, the security architecture of GSM networks can be viewed as a subset of that of UMTS networks.

In Figure 2.1, we have six different entities, each of which can exist in many instances in a real network. There is a fair amount of symmetry in this security architecture. We continue by describing UMTS security architecture via these symmetries, five of them in total.

The **first symmetry** is on the generation of the keys and other authentication data.

Beginning from the top of Figure 2.1, the Authentication Center (AuC) is the network element responsible for storing the permanent cryptographic keys bound to the subscription and computing the session keys and the authentication data. The AuC is usually implemented as an integral part of Home Location Register (HLR) or in latter releases of the 3GPP specifications Home Subscriber Server (HSS) which is the main operator database, containing all subscriber data, including information about the whereabouts of the subscriber. Often the HSS is implemented as an HLR with extended functionality.

In the AuC calculations, the permanent key for a subscriber is denoted by K. Another security-relevant parameter is the sequence number SQN_{AuC}. As the name already indicates, sequence numbers are constantly changing, in contrast to the

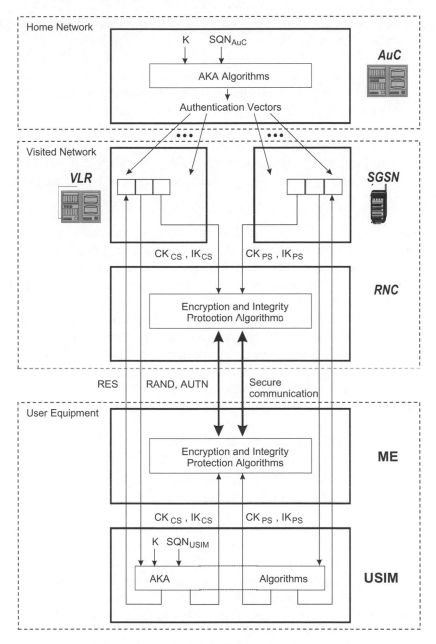

Figure 2.1. Summary of the UMTS access security features

master key K. AKA algorithms are used to create a UMTS Authentication Vector (AV), containing a quintuple of:

- RAND: random 128-bit number generated by AuC
- XRES: derived by a one-way function from K, RAND and SQN_{AuC}
- CK: a session key that is also derived by a one way function from the same input parameters
- IK: another session key derived similarly
- AUTN: an authentication token containing SQN_{AuC} in encrypted form, an administrative field AMF and a message authentication code (MAC) that protects the integrity of the AV.

Moving next to the bottom, we find the counterpart of the AuC on the user side: UMTS Subscriber Identity Module (USIM). This is an application residing in a special kind of smart card, the Universal Integrated Circuit Card (UICC). The technical representation of the subscription is the subscriber-specific shared master key K. Thus, the master key K is stored permanently in the USIM, as well as in the AuC. If two of the AV parameters, RAND and AUTN, are given as input parameters to the USIM, very similar set of algorithms, compared to those used in the AuC, are used to calculate CK, IK and another parameter RES that should be identical to XRES if everything has been done correctly. The subscriber's device, called Mobile Equipment (ME) is capable of interfacing with the UICC.

Next, we describe the **second symmetry**: this is related to the architecture of the UMTS as a whole.
There are two different domains that act independently in many respects. In Figure 2.1, there are two elements on the second layer from the top:

- Visitor Location Register (VLR) belongs to the Circuit-Switched (CS) domain while the
- Serving GPRS Support Node (SGSN) belongs to the Packet-Switched (PS) domain.

The main service provided by the CS domain (resp. PS domain) is the voice calls (resp. data transfer). Both VLR and SGSN fetch authentication vectors from the HLR/AuC node whenever a subscriber has attached to them and there is need for authentication of the subscriber. The session keys in these AVs for each domain are marked by the subscripts 'CS' and 'PS'.

Our **third symmetry** is related to the authentication: the network wants to authenticate the subscriber. On the other hand, it is also in the subscriber's interest to authenticate the network to avoid, for example, false base-station attacks.

For each of the domains, the authentication of the subscriber is carried out in an identical manner. First, VLR (resp. SGSN) sends the two parameters RAND and AUTN to the USIM. (There are several network elements on the path but they are irrelevant in this context.) The USIM checks first that the Message Authentication Code (MAC) in the AUTN is correct. Next, the USIM checks that it has not seen the sequence number hidden in the AUTN before. If these two checks lead to a positive end result, it is concluded, first, that the parameters have truly been generated in the AuC and, second, that the parameters are used first time in this particular authentication (this is usually called protection against replay attacks). What follows is that the USIM sends the calculated response parameter RES back to the VLR (resp. SGSN). A comparison with the parameter XRES concludes the authentication procedure.

The generated session keys lead us to the **fourth symmetry**.

The VLR (resp. SGSN) transfers the session keys CK and IK to the Radio Network Controller (RNC). Similarly, the USIM provides identical keys to the ME. Note that in the case of successful authentication, the parameters RES and XRES have found to be identical and there is a great probability that the session keys on both sides are identical as well.

Our **fifth symmetry** is between the usage of the session keys CK and IK.

Indeed, CK is used for confidentiality protection purposes while the IK is used for integrity protection purposes. In addition to these session keys, appropriate cryptographic algorithms are needed here. In UMTS Release 99 specifications, one confidentiality algorithm and one integrity algorithm were defined [TS35.201]. Both algorithms are based on the same block cipher cryptographic algorithm called KASUMI [TS35.202]. KASUMI uses 128-bit keys and a block size of 64 bits. One difference between the confidentiality and integrity protection mechanisms is that the former is applied to both user traffic and signalling traffic, while the latter is only used for signalling traffic.

Next, we briefly list the main differences between the Global System for Mobile communication (GSM) access security architecture and the UMTS access security architecture.

Our first symmetry exists also for the GSM access parameters. Instead of the 5-parameter long AV, only three parameters are in use:

- random challenge (RAND),
- (expected) response (now called SRES) and
- the cipher key (now called Kc).

The counterpart of the AuC on the user side is called the Subscriber Identity Module (SIM). The SIM is a smart card itself. The second symmetry is visible also in GSM. Actually, the UMTS core network architecture is inherited from the GSM and its packet-switched extension, GPRS. The third symmetry does not exist in GSM / GPRS: in these systems, only authentication of the subscriber is carried out. The fourth symmetry is visible in GSM and GPRS but there are differences, both compared to the UMTS system and also between the solutions of GSM and GPRS.

In GSM, the cipher key Kc is transferred all the way to the Base Station (BS) while, on the other hand, in GPRS, the cipher key stays in the SGSN, it is only transferred to the ciphering function. The fifth symmetry is not applicable to GSM since there is no separate integrity protection mechanism in GSM (or in GPRS).

The ciphering algorithms are also different in GSM / GPRS compared to those used in UMTS. The most notable difference is length of the key: the key Kc of GSM / GPRS has a length of 64 bits. The specifications of the first two versions of both GSM and GPRS encryption algorithms (A5/1 and A5/2 for GSM, GPRS Encryption Algorithm (GEA) versions, GEA1 and GEA2, for GPRS) have been kept private and available only to the mobile operators and to vendors of the infrastructure elements and terminals. There are also KASUMI-based public version A5/3 and GEA3 available nowadays which address some shortcomings of the previous algorithms.

Some other security features of UMTS and GSM / GPRS relevant for the rest of the book are briefly described in the balance of this section.

The user identity is protected by the usage of temporary identities. As soon as the permanent identity of the subscriber, the IMSI (International Mobile Subscriber Identity), is known to the network (i.e., VLR or SGSN), the network begins to use a temporary identity, TMSI (Temporary Mobile Subscriber Identity) instead. The TMSI is sent to the User Equipment (UE) over the encrypted channel. The UE is a mobile terminal with a smart card inserted. Whenever the TMSI has been used by the UE to, for example, set up a call, receiving a call or location update, the network provides a new TMSI for the UE. The TMSI mechanism does not give protection of user identity confidentiality against *active* attacks because the attacker can always pretend to be a network element that has lost the TMSI, and thus, requests the permanent identity from the UE. But the TMSI protects against passive attacks like eavesdropping.

The session keys CK, IK and Kc may be used for several sessions. In other words, authentication is not required for each session. There is no lifetime associated to the keys, though. In UMTS, there is a possibility to restrict how much of data is protected by the same session key but this amount is typically very high and the protection is intended for extremely long sessions only. The reason for omission of the lifetime is that in case a subscription itself expires, information about this is carried to the serving network that can technically then simply cut down the session.

In Figure 2.2 we show the information flow related to the AKA protocol.

Figure 2.2 shows the exchanges involved in the UMTS AKA protocol. Subscribers access services by connecting to a serving network. The serving network depends on the subscriber's present location: when he is at home, the serving network is the same as the home network; when he is travelling elsewhere, it is usually a network of a different operator who has a roaming agreement with the home network operator.

The subscriber's USIM and the Home Network share a long-term key K and maintain a running sequence number SQN.

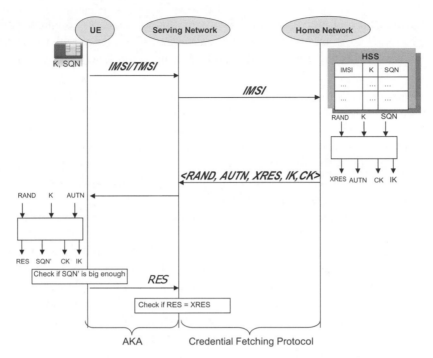

Figure 2.2. UMTS authentication and key agreement (AKA) protocol

1. The authentication procedure is triggered by the UE sending its permanent identity to a server in the serving network (home or visited network). The first time UE connects to a serving network it has to use its true IMSI as the identifier. But on successful authentication, the serving network may provide a temporary identifier TMSI. The serving network maintains the mapping of TMSIs to IMSIs.

2. The serving network forwards the IMSI to the home network which chooses a random challenge RAND and looks up K and the current SQN for IMSI. RAND, K and SQN are fed into a set of cryptographic algorithms to produce four values: an authenticator AUTN to prove that RAND was sent by the home network, an expected response XRES for the given RAND and two cryptographic session keys CK and IK to be used to protect confidentiality and integrity of subsequent communication between the UE and the serving network.

3. The Home Network sends these four values along with the corresponding RAND to the serving network.

4. The serving network forwards RAND and AUTN to UE.

5. The UE's USIM takes RAND, AUTN and K as input to the same set of cryptographic algorithms as in step 2 to produce a response RES, the sequence number SQN' used by the Home Network, and the two keys IK and CK.

6. The USIM checks if the value SQN' is big enough given its own sequence number value SQN. If so, it accepts the authentication and sends RES back to the serving network. If RES is the same as XRES, the serving network accepts the authentication as well.

Thus, AKA provides mutual authentication between the subscriber and the network operator. It also results in temporary session keys shared between the UE and the serving network. The above is an extremely simplified overview of the UMTS security architecture intended to present only those aspects relevant for eliciting the requirements for GAA. Readers interested in a comprehensive description of UMTS security architecture should consult [Niemi07].

In normal AKA usage, the UE sends and receives AKA message over its radio interfaces. But there is nothing inherent in AKA that binds it to UMTS radio layer. Therefore, in course of time, two new transports for AKA signalling were specified.

The first is HTTP Digest AKA [RFC3310]. HTTP Digest AKA defines a new 'algorithm' for the HTTP Digest authentication scheme [RFC2617] by mapping AKA parameters to the information elements required by HTTP Digest authentication. Essentially, HTTP Digest AKA encodes the RAND and AUTN parameters into the nonce field of an HTTP Unauthorized response from the server; the client extracts RAND and AUTN, applies the normal AKA processing, and uses the resulting RES parameter of AKA as the 'password' for HTTP Digest Authentication, thereby turning HTTP Digest into a one-time password scheme. The resulting IK and CK can be used to protect the subsequent communication between the client and the server. Usually, HTTP Digest authentication is used in conjunction with a server-authenticated TLS tunnel so that subsequent communication between the server and the client is protected by the TLS session key. However, HTTP Digest AKA authentication **must not** be used in this manner because an attacker pretending to be a serving network could easily get RES from a UE using regular AKA (see [Asokan05] for details of this type of man-in-the-middle attacks).

The second alternative transport for AKA is over Extensible Authentication Protocol (EAP) defined in the EAP-AKA RFC [RFC4187]. This specification defines a new EAP authentication method to transport AKA signalling. This allows AKA to be used in applications that support EAP, usually for access authentication, e.g., in WiFi networks.

2.1.2 Issues in Securing Services with Radio Layer Security

Many service providers have concluded that the basic security level provided by the radio layer (i.e., the link layer) is sufficient for their services that are provided on higher layers. From the point of view of confidentiality, this conclusion may be well justified in many cases. Indeed, the radio interface is the one that is most vulnerable to the eavesdroppers. Encryption on the radio layer provides protection against such attackers for all data in higher layers as well. On the other hand, from the point of view of authentication, the situation is more complicated.

This is because the well-accepted layered approach in communications (e.g., OSI 7-layer model, IETF 5-layer model) does not fit optimally with authentication. The reason for this stems from the following facts:

- Authentication is always done in relation to a certain identity, i.e., verifying that identity.
- The layers use different identities (e.g., MAC address, IP address, Session Initiation Protocol (SIP) identity).

Therefore, we have to choose one of the following approaches:

(a) execute authentication in many layers (which obviously adds complexity to the system);
(b) let some identities be unauthenticated (which implies that certain threats cannot be addressed at all); and
(c) bind different identities together.

The last solution has also drawbacks:

- The binding has to be done securely which adds the need for a new security feature to the system. Also, the binding may change frequently and cause quite some load to the systems. Network Address Translation (NAT)[1] servers may cause additional complexity if used together with identity binding for IP addresses.
- The binding causes a violation against the core idea of the layered communication model.

In 3GPP systems, all of these approaches are in use, e.g., in IP Multimedia Core Network Subsystem (IMS):

- Full IMS with IP Multimedia Subsystem Identity Module (ISIM) application on the UICC is an example of approach (a) [TS31.103], [TS33.203]. This is a very secure and scalable approach; the largest drawback is that the penetration of ISIM applications on the UICC is still limited and the network support for this is not that wide.
- HTTP digest authentication over insecure access. This has been specified as an optional mechanism for fixed networks in ETSI TISPAN. This is an easy, but not very secure, approach. It is an example of approach (b) in the sense that access-level identities are not assumed to be authenticated.

---

[1] http://en.wikipedia.org/wiki/Network_address_translation

- Early IMS security [TR33.978], i.e., IP address binding is an example of approach (c). This approach works, but has some scalability issues, when a large range of users using a large set of service (potentially with NATs deployed in the system). Also IP address spoofing may become an issue with this approach. Frequent IP address changes cause additional complexity and load to this approach.
- Network Attachment Subsystem (NASS)-IMS bundled authentication, i.e., binding to the calling line identifier [TS33.203] is another example of approach (c). This approach works well enough for simple use cases, but does not support mobility. If mobility or several users behind one line should be supported, then additional features or functionalities are required.

2.2 General-Purpose Approaches to Authentication and Key Management

2.2.1 Public Key Infrastructure (PKI)

Public key cryptography was invented in the mid-1970s [Diffie76]. Two groundbreaking concepts were introduced:

(1) Separation of keys used for encryption and decryption. As a consequence, one of the keys could be made public without revealing the other key.
(2) Digital signature. It became possible to verify authenticity of a completely digital document in a way that also enabled unforgeability: only the source could have created the document with the digital signature.

It was first thought that concept (1) would make key management of a big system very easy. Every entity in the system could have its own private key, stored and protected in one single entity. The corresponding public key could be made available for every other entity in the system by including it in an appropriate database. Soon it was observed, however, that when the system becomes even bigger and less centrally managed, the database solution for managing public keys does not scale anymore. It becomes difficult to guarantee access to the database in a secure manner, and, on the other hand, it becomes, more and more difficult to guarantee that the public keys in the database are authentic themselves.

Fortunately, concept (2) provided solutions to these problems. Public keys could be digitally signed by an authority, thus creating a *certificate*. Then there is no need to have a secure online access to a centralized authentic database: if you want to send a message to a person, you could find the digitally signed certificate from whatever source (e.g., from the Internet).

However, the issue is not so straightforward: verifying of digital signatures itself requires authentic keys. In the setting described above, the sender of the message has to have access to the authentic verification key of the authority. Otherwise, it is not possible to check the digital signature in the certificate, and in consequence, it would not be possible to trust the public key in the certificate.

It may seem that we are back to square one: in order to be able to fetch securely a public key of the receiver, we first have to fetch securely the public key of the authority. Luckily, the number of authorities is much smaller than the number of all entities in the system, and therefore, we have converted a big scale issue to a much smaller scale issue. As a solution to the remaining smaller scale issue, we iterate the same process: we have a master authority that signs the public signature key of the lower-level authority. In this way, we have created a *chain of certificates*.

Our wish would be that the number of master authorities would be so small that the corresponding public keys could be installed to the system entities securely in some direct manner, for instance, at the same time when the public key cryptographic functions are installed. In some cases, our wish does not materialize, and we have to add further layers of authorities to the system. All these authorities constitute the public key infrastructure. In addition to *Certificate Authorities* (CA), the PKI typically needs also *Registration Authorities* (RA) for the purpose of introducing new entities to the system: authenticating users physically by the means of, e.g., checking their identity cards or driver's licence.

The PKI constitutes a generic-purpose authentication and key management system, and in principle, it would suit for mobile systems as well. One challenge with the mobile systems is the fact that these systems tend to be global, putting up the requirement for a global PKI. Building a global infrastructure of any sort is a big effort, not only technically and financially but also politically. (See Figure 2.3.)

One of the biggest existing use case for PKI stems from the use of certificates in the Transport Layer Security (TLS) protocol [RFC2246]. It has been possible to issue certificates to a large number of companies and servers. What is still largely missing at the time of writing is a wide base of *client* certificates. TLS supports client

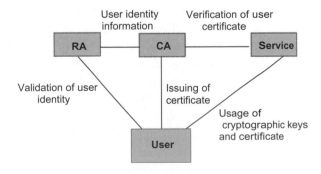

Figure 2.3.   An example of Public Key Infrastructure (PKI)

certificates but, on the other hand, a server certificate is sufficient to put up a TLS session. In a later section, we describe how GAA could be used to support client or *subscriber* certificates.

In the wireless area, the largest initiative to build a PKI has been *Wireless PKI (WPKI),* specified originally in the Wireless Application Protocol (WAP) Forum, and later continued in the Open Mobile Alliance (OMA) (see [OMAWPKI]). An important element of WPKI is *WIM (Wireless Identity Module)* (see [OMAWIM]).

2.2.2 Passwords

The dominant approach in authentication of applications in the Internet is to use passwords. Each user has a username and a password that is assumed to be known only by the user. The password could be changed by regular time intervals but is also often the case that the password is never changed. From security point of view, it would be best to use *one-time passwords*. Obviously, they have the disadvantage that there has to be a way of securely transferring one password per session to the user. One-time passwords are typically used in certain security-critical applications where authentication is needed typically in the frequency of daily or more seldom. Internet banking is an example of such application. Of course, user cannot memorize one-time passwords, which implies that the list of unused one-time passwords has to be stored securely while still kept handily available.

If the user would need to send over the password as such every time she wants to access the service, an eavesdropper anyway on the path from the user to the server would have plenty of opportunities to snatch the password. That would enable the attacker to masquerade as the user against the server. Obviously, in the wireless environment, an attacker would have easy access to the path but there are typically also vulnerable points on the network interfaces. In order to protect against this kind of attacks, what is usually sent over is not the password itself but something derived from it. Some time-varying or random parameter needs to be added to the derivation, otherwise, the eavesdropper would just need to capture the data derived from the password and use that for masquerade purposes.

The dominant mechanism user for username-password is the HTML form-based authentication where the web page itself contain username and password fields. The dominant protocol used for username-password mechanism in the Internet is called HTTP Basic and Digest [RFC2617]. The difference between the two is that HTML forms are carried in the HTTP payload while the HTTP Basic and Digest use the HTTP headers to carry the username-password data. With HTML forms, the user types in the username and password and they are sent to the server. They are typically protected using https, i.e., SSL or TLS [RFC2246], in which the HTTP communication is conducted via encrypted secure channel. The difference between HTTP Basic and Digest authentication is that in Basic, the username and password are sent in the clear to the server (or actually base64 encoded), and Digest uses message digests (i.e., MD5) to protect the password. Moreover, HTTP Digest authentication is of 'challenge-response' type. The server is sending over a challenge in the form of parameter called 'nonce'. The response from the user must contain a parameter 'response' that contains a hash function value derived from the username, password and nonce, together with some other parameters. It is worth noting that the server does not have to store the password as such or actually it does not even have to know it. It is enough that an intermediate value derived as well by a hash function from the password and the username (together with some other parameters).

Later in the book we show how GBA could be used to support HTTP Digest.

2.2.3 Kerberos

There are also key general-purpose key management systems based on shared secret keys. One of the most used systems is Kerberos, developed by MIT in the 1980s (see [RFC4120]). It is supported by, e.g., all recent versions of Windows operating system. Kerberos is based on a *Key Distribution Center* (KDC) that shares a permanent master key with each user of the system. It is also assumed that well-synchronized clocks are available for each user. These are needed to be able to provide time-stamped *tickets* for users. These tickets can then be used as authentication tokens in order to grant access to services.

One disadvantage of Kerberos stems from its dependency on the KDC that acts as a centralized, trusted third party. This introduces similar issues as the centralized CA approach. Fast-expiring tickets put high availability requirements on the KDC. Availability of synchronized clocks is another constraint, especially in the mobile environment.

2.2.4 Radio Layer and General Purpose Security Mechanisms

As hinted in earlier sections, before GAA, there had already been earlier attempts to utilize the existing access security architecture for other purposes than cellular radio access.

One approach is to use *short message service* (SMS) as an independent channel to carry authentication data. For example, in order to get access to a service in the Internet, a user could first send a request over SMS to the number provided by the service. By checking the phone number, the service would be able to identify the user. As a next step, a one-time password could be sent to the UE, to be subsequently used when accessing the service over the Internet, using, e.g., HTTP Digest protocol. Note here that in case an attacker could spoof its phone number in the SMS, it would still not receive the password because it would be routed to another UE (if anywhere). It is also worth noting that the actual service could be accessed either by the same UE over the cellular network or by another machine, e.g., a home PC, over a fixed broadband network.

Another approach was introduced for the purpose of securing access to Wireless Local Area Networks (WLANs). The original idea was to take the SIM-based authentication protocol and embed that in the WLAN protocols somehow. In this approach, Extensible Authentication Protocol (EAP) [RFC3748] plays a major role. It is a protocol that supports various authentication methods and it can be run on top of various link layer protocols, e.g., on top of IEEE 802 protocols. The SIM authentication mechanism has been introduced to EAP as a specific EAP method, called EAP-SIM, in [RFC4186]. Similarly, UMTS AKA has been specified as an EAP method, called EAP-AKA, in [RFC4187].

Another approach was introduced for securing access to IMS. For the purpose of the so-called full IMS security solution [TS33.203], the AKA protocol was embedded into HTTP Digest protocol as an extension [RFC 3310]. This mechanism is also the cornerstone of GAA.

In the full IMS security system keys generated during the AKA run are used for securing SIP signalling messages. This is done by deriving IPsec Security Associations from the AKA keys and some other parameters. Then IPsec is used to protect the confidentiality and integrity of signalling messages between the terminal and the first IMS entity, an SIP server called Proxy CSCF.

2.3 Requirements for GAA

So far in this chapter, we have examined existing approaches for authentication. As we pointed out in Chapter 1, there are many compelling reasons for bootstrapping credentials from existing, widely used, authentication infrastructures. Before we go on to describe the design of the GAA, let us consider desirable characteristics of a general-purpose AKA architecture, especially when it is bootstrapped from an existing authentication infrastructure.

Generality: Many different types of applications and services must be able to use GAA for authentication. For simplicity, let us call these GAA applications. For example, broadcast Mobile TV with smart card profile authentication that is specified by OMA BCAST group [OMASC] is a GAA application.

Application separation: Security guarantees of one GAA application must not depend on the correct behaviour of other GAA applications. For example, suppose there is a flaw in the design or implementation of a GAA application protocol, and this results in the exposure of secret data used by that application, this must have no effect on the secret data used by other GAA applications. Similarly, a misconfigured application server or client may result in the exposure of secret data used by that instance. This must have no effect on other servers or clients even if they are using the same GAA application.

Access independence: Use of GAA must not depend on a particular access technology. For example, earlier in this chapter, we saw how SMS messages may be used for bootstrapping authentication. This is an access-specific mechanism since SMS messages can be sent and received only when a device is connected to the cellular access network.

Reuse: In order to maximize the benefits of bootstrapping, GAA should reuse existing protocols and infrastructure wherever possible. It should also be extensible, in that sense, that future applications can reuse GAA easily.

Protection of original infrastructure: A flaw in the design, implementation or configuration of GAA or any GAA application must not harm security and operation of the UMTS security infrastructure itself.

Control by home network: The home network operator is the entity that has a business relationship with the subscriber. Therefore, the home network operator should be able to set the policy what GAA applications are available to a given subscriber.

In Chapter 3, we will discuss how those requirements are realized in practice.

References

[Asokan05]   N. Asokan, Valtteri Niemi and Kaisa Nyberg, *Man-in-the-middle in tunneled authentication protocols*, Proceedings of 11th Cambridge Workshop on Security Protocols, Springer Lecture Notes in Computer Science 3364 (2005), 28–41.

[Diffie76]   Whit Diffie, Martin Hellman, *New Directions in Cryptography*, In IEEE Transactions on Information Theory, 22:6 (644–654), November 1976. Available at http://ieeexplore. ieee.org/xpls/abs_all.jsp?arnumber=1055638

[Niemi07]   Valtteri Niemi, Kaisa Nyberg, *UMTS Security*, Wiley & Sons (2003).

[OMASC]   Open Mobile Alliance (OMA), OMA-TS-BCAST_SvcCntProtection-V1_ 0-20070529-C, *Service and Content Protection for Mobile Broadcast Services Specification*, Candidate Version 1.0 – September 2007, http://www.openmobilealliance.org/ release_program/bcast_v1_0.html

[OMAWIM]   Open Mobile Alliance (OMA), OMA-WAP-WIM-V1_1-20021024-C, *Wireless Identity Module (WIM), Part: Security*, Version 1.1 – October 2002. Available at http://www. openmobilealliance.org/release_program/wpki_v10.html

[OMAWPKI]   Open Mobile Alliance (OMA), OMA-WPKI-V1_0-20040615-C, *OMA Wireless Public Key Infrastructure*, Version 1.0 – June 2004. Available at http://www.openmobilealliance.org/release_program/wpki_v10.html

[RFC2246]   Internet Engineering Task Force (IETF), *The TLS Protocol Version 1.0*, RFC 2246, January 1999. Available at http://www.ietf.org/rfc/rfc2246.txt

[RFC2617]   Internet Engineering Task Force (IETF), *HTTP Authentication: Basic and Digest Access Authentication*, RFC 2617, June 1999. Available at http://www.ietf.org/rfc/rfc2617.txt

[RFC3310]   Internet Engineering Task Force (IETF), *Hypertext Transfer Protocol (HTTP) Digest Authentication Using Authentication and Key Agreement (AKA)*, RFC 3310, September 2002. Available at http://www.ietf.org/rfc/rfc3310.txt

[RFC4120]   Internet Engineering Task Force (IETF), *The Kerberos Network Authentication Service (v5)*, RFC 4120, July 2005. Available at http://www.ietf.org/rfc/rfc4120.txt

[RFC4186]   Internet Engineering Task Force (IETF), *Extensible Authentication Protocol Method for Global System for Mobile Communications (GSM) Subscriber Identity Modules (EAP-SIM)*, RFC 4186, January 2006. Available at http://www.ietf.org/rfc/rfc4186.txt

[RFC4187]   Internet Engineering Task Force (IETF), *Extensible Authentication Protocol Method for 3rd Generation Authentication and Key Agreement (EAP-AKA)*, RFC 4187, January 2006. Available at http://www.ietf.org/rfc/rfc4187.txt

[TR33.978]   3rd Generation Partnership Project (3GPP), Technical Report TR 33.978, *Security Aspects of Early IP Multimedia Subsystem (IMS)*, Version 7.0.0 (2007). Available at http://www.3gpp.org/

[TS31.103]   3rd Generation Partnership Project (3GPP), Technical Specification TS 31.103, *Characteristics of the IP Multimedia Services Identity Module (ISIM) application*, Version 7.1.0 (2006). Available at http://www.3gpp.org/

[TS33.203]   3rd Generation Partnership Project (3GPP), Technical Specification TS 33.203, *3G Security; Access Security for IP-based Services*, Version 8.1.0 (2007). Available at http://www.3gpp.org/

[TS33.102]   3rd Generation Partnership Project (3GPP), Technical Specification TS 33.102, *3G Security, Security Architecture*, Version 7.1.0 (2006). Available at http://www.3gpp.org/

[TS35.201]   3rd Generation Partnership Project (3GPP), Technical Specification TS 55.201, *Specification of the 3GPP confidentiality and integrity algorithms; Document 1: f8 and f9 specification*, Version 7.0.0 (2007). Available at http://www.3gpp.org/

[TS35.202]   3rd Generation Partnership Project (3GPP), Technical Specification TS 55.202, *Specification of the 3GPP confidentiality and integrity algorithms; Document 2: Kasumi specification*, Version 7.0.0 (2007). Available at http://www.3gpp.org/

3

Generic Authentication Architecture

3.1 Overview of Generic Authentication Architecture

In this chapter, we will present the overall picture of the design of GAA. We start by describing how the requirements identified in Section 2.3 influenced important design decisions. An in-depth description of the technical details appears in the subsequent sections of this chapter.

3.1.1 Rationales for Design Decisions

The first design decision is the type of credentials that should be bootstrapped from the parent infrastructure.

The **generality** requirement states that the bootstrapped credentials should be usable with a wide variety of applications. In GAA, the bootstrapped credential is in the form of a transaction identifier and a temporary shared secret key. This part of GAA is called as the Generic Bootstrapping Architecture (GBA). In GBA, a temporary shared session key is obtained by running the authentication protocol in the parent infrastructure. In the case of UMTS networks, this protocol is the AKA protocol (see [TS33.102] or [RFC3310]).

This session key can be used directly in any application which uses username/ password style of authentication. Alternately, it can be used to initialize other types of credentials. For example, it can be used to enrol a public key of the subscriber to an operator-run Public Key Infrastructure (PKI). The PKI can then issue a subscriber certificate to be used with applications that support public-key authentication. Thus,

Cellular Authentication for Mobile and Internet Services
Silke Holtmanns, Valtteri Niemi, Philip Ginzboorg, Pekka Laitinen and N. Asokan
© 2008 John Wiley & Sons, Ltd

GBA constitutes the foundation of GAA, which consists of GBA and several other forms of application authentication, built on top of GBA. Together they help meet the generality requirement.

The **access independence** requirement implies that the bootstrapping procedure must be the same regardless of the type of network currently being used by the bootstrapping device. As we saw earlier, there are currently two ways to transport the messages for UMTS AKA in an access-independent manner: using Digest AKA [RFC3310] or using the Extensible Authentication Protocol EAP-AKA [RFC4187]. GAA uses Digest AKA within HTTP when bootstrapping from the UMTS infrastructure because EAP is typically used for access authentication rather than service authentication.

The discussion above already shows that also the requirement of **reuse** has been satisfied with GAA.

To satisfy the **protection of original infrastructure** requirement, the session keys resulting from original authentication protocol are not used directly to secure GAA applications. Instead, these session keys are used as the GAA master session key. The shared session key for GAA applications are obtained by applying a *key diversification* function to the GAA master session key. The key diversification function is *one-way*: it is easy to compute the GAA application key from the AKA session keys, but it is infeasible to do the reverse. This way, even if the GAA application key is exposed due to a design or configuration flaw in a GAA application, it cannot be used to attack original infrastructure.

Similarly, the GAA transaction identifier is independent of identifiers in the original infrastructure, such as UMTS subscriber identities like the International Mobile Subscriber Identity (IMSI) or Mobile Station International ISDN Number (MSISDN). Moreover, the average extra load on the original infrastructure from GAA application requests is controlled by the lifetime of bootstrapped keys. If a GAA application uses the bootstrapped keys only for the indicated lifetime, it can automatically detect when the subscriber's keys are revoked in the original infrastructure. It is entirely up to the GAA application to decide whether or not to use the bootstrapped keys only for the indicated lifetime.

In order to achieve **application separation**, we cannot use a single shared session key for all GAA applications. Instead, each GAA application must be given mutually independent key material. This *key separation* is implemented by including the application server's identity and application protocol's identity as an input parameter to the key diversification function. As a result, the same subscriber's device will end up using a different GAA application key with each different GAA application. Furthermore, key diversification guarantees that one GAA application key cannot be used to derive another GAA application key.

Finally, a number of mechanisms are used for achieving **home control**. The GAA master session key and all the associated application-specific keys have limited lifetimes specified by the home operator. The home operator also maintains application-specific user profile for each subscriber, which specifies what information about a subscriber is released to an application server, and to configure the type of service that an application server is allowed to provide to a subscriber.

3.1.2 A Bird's Eye View of GAA

Figure 3.1 shows a schematic view of the basic GAA. There are three different general network functions in GAA:

- The **Home Server (HS)** is the subscriber database and contains the long-term subscriber key for each subscriber. It is a standard function in cellular networks. For example, in UMTS networks (according to latest set of specifications), it is known as the HSS. In GSM networks (and older UMTS networks), the HS is known as Home Location Register (HLR).
- **Bootstrapping Server Function (BSF)** is a new network function introduced in GAA. It facilitates the use of AKA to bootstrap a new GAA master session key.
- The server functionality of each GAA server application is a **Network Application Function (NAF)**.

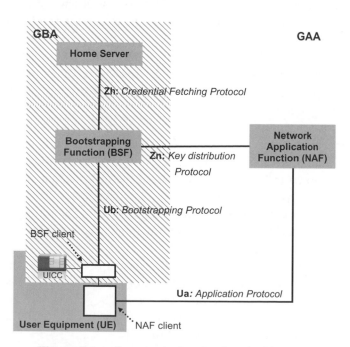

Figure 3.1.   Generic Authentication Architecture

A network function is an abstract (or logical) construct. Each network function may be implemented in a separate physical network elements. But it is possible for some network functions to be co-located in the same network element. For more details on each network element, see Section 3.2.1.

The client device, also known as the **User Equipment (UE)**, contains the client-side functionality. The long-term subscriber keys are contained in subscriber identity modules. For example, in UMTS USIM, application on a smart card is used for enabling packet- and circuit-switched access. Typically, the subscriber identity modules are housed in the subscriber's smart card, known as Universal Integrated Circuit Card (UICC). The BSF client is the entity in UE that participates in bootstrapping. It interacts with the BSF on the network and the subscriber identity module in the UE. The UE will also have one or more NAF clients, which obtain an application-specific bootstrapped key from the BSF client and use it to secure the application protocol. A NAF client is an application-specific software element in the device, e.g., streaming application, browser application, etc. The NAF server obtains the same key from the BSF server.

The protocols for the interactions between two entities are specified by the **interface** (also called reference point) between them. There are four such interfaces in the basic GAA. We explain the interfaces by describing the procedures in which they are used: bootstrapping of a shared key and the use of that key. Strictly speaking, there is a difference between the notions of 'interface' and 'reference point', i.e., the former refers to a boundary between two physical entities while the latter refers to a connection point between two logical functions. But in the sequel, these terms are used interchangeably.

The first procedure is GAA bootstrapping. This is the process by which the AKA protocol in the parent infrastructure is used to set up a GAA master session key between the UE and the BSF. The starting point of bootstrapping is that the UE and Home Server share a long-term key and can use it to run an AKA protocol specified for the parent infrastructure. The exchanges between UE and BSF during the bootstrapping procedure are specified by the Ub interface. The exchanges between BSF and the Home Server are specified by the Zh interface. The typical GAA bootstrapping procedure shown in Figure 3.2 consists of the following steps:

1. The BSF client in UE initiates bootstrapping by sending a request to the BSF with the identity by which the subscriber is known in the original infrastructure. Thus, is done over the Ub interface.
2. This triggers a run of the authentication protocol between UE and HS with the BSF acting as the intermediary. At the end of this run, UE and BSF obtain a set of shared session keys. The GAA master session key Ks is derived from this set. In the case of UMTS AKA, the session keys are known as IK and CK. Ks is simply the concatenation of the IK and the CK.

3. The BSF also receives a set of user profiles from the HS over Zh interface if the HS is an HSS. Each GAA application may have a user profile.
4. BSF constructs a transaction identifier B-TID and stores B-TID, Ks and the user profiles in its database. It also chooses a key lifetime according to its local policy.
5. BSF sends B-TID and the key lifetime to the UE.
6. UE stores B-TID, Ks and key lifetime.

At this point, the bootstrapping is complete. Both UE and BSF share a temporary GAA master session key Ks and a transaction identifier B-TID that can be used to identify Ks. It should be noted that the Ks itself is not bound to a particular application or NAF.

The second procedure is the usage of bootstrapped keys. This is the process by which UE can use the bootstrapped keys to secure its exchanges in an application protocol with a NAF acting as the application server. The exchanges between UE and NAF in the application protocol, during the bootstrapping usage procedure, are

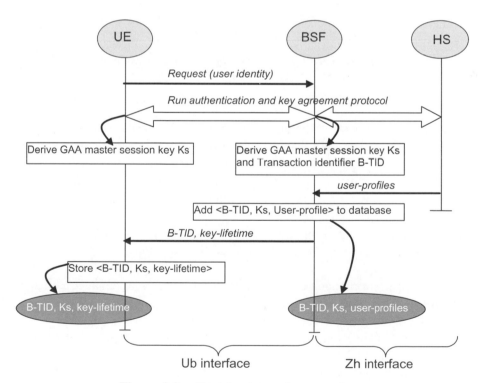

Figure 3.2.   GAA bootstrapping procedure

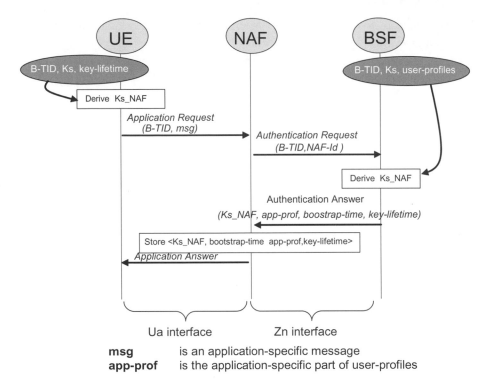

Figure 3.3.   GAA Bootstrapping usage procedure

specified by the Ua interface. The exchanges between NAF and the BSF are specified by the Zn interface. If the NAF and the BSF reside in different networks, an intermediate proxy (Zn proxy) is needed to communicate to the BSF. The interface between the NAF and the Zn proxy is the Zn interface. The interface between the Zn proxy and the home-BSF is called Zn' interface (spelled 'Zn prime'). For more details on Zn proxy, see Section 3.5.5.

Figure 3.3 illustrates the typical form of bootstrapping usage (for other cases see the use cases in chapter 4). It consists of the following steps:

1. Once the application-specific NAF client in UE decides to engage in an application protocol run with a specific NAF, it derives an application-specific session key Ks_NAF from the GAA master session key Ks.
2. UE starts the application protocol by sending a request which contains the key identifier, i.e., B-TID over Ua interface.
3. NAF forwards B-TID and its own identifier NAF-Id to the BSF over Zn interface (if it resides in a visited network then Zn' and Zn interface are used together).

4. BSF can now look up its database for the entry corresponding to the B-TID. It first decides whether the NAF is authorized to receive application-specific keys for this particular subscriber. This is determined according to local policy and the policy in user profiles obtained from the HS.
5. If the decision is positive, BSF derives the application-specific session key Ks_NAF from Ks. It also selects the subset of user profiles that the NAF is authorized to receive.
6. BSF sends Ks_NAF, the application-specific subset of user profiles and other relevant information to the NAF over Zn interface.
7. NAF stores the received information. At this point, UE and NAF share a temporary GAA application key, i.e., the application-specific session key Ks_NAF, and can use it to protect subsequent exchanges in the application protocol.
8. NAF replies to UE using the application protocol and the application-specific session key Ks_NAF.

The key derivation procedure used in steps one and five has two properties as shown in Figure 3.4. First, it uses a one-way function so that while it is easy to compute Ks_NAF from Ks, it is extremely difficult to compute Ks given Ks_NAF. Second, an identifier for the NAF as well as session-specific information are used as inputs to the procedure so that the Ks_NAF keys for different applications and different sessions will be mutually independent. In Sections 3.2 and 3.3, we will see exactly how key derivation is done.

So far we have gained a conceptual overview of GAA. The details vary depending on what the original infrastructure is (e.g., UMTS or GSM), where the resulting application-specific keys are to be stored and used in the UE (inside the UICC or in the mobile terminal). In addition to the primary entities and interfaces we have seen so far, there are other components of GAA. Now we are ready to look at the technical details of GAA.

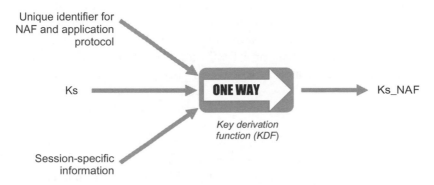

Figure 3.4. Deriving application-specific keys with key separation

3.2 Foundations of GAA

In the previous section, the GAA architecture has been outlined and an overview of GAA was provided. In this section, we present the detailed description of GAA that has been specified in 3GPP.

3.2.1 Architectural Elements of GAA

The basic GAA architecture in a 3G or 2G network consists of four different components (see also more general description in Section 3.1.2):

- **HSS** or **HLR** that contains the subscriber data, and from GAA perspective it contains the long-term master key K that is shared between the terminal and the operator and GBA User Security Settings (GUSS) data. In more general terms, the HSS and HLR are described as HS in Section 3.1.2.
- **BSF** that serves as a credential server. The BSF is the centrepiece of GAA by establishing the GAA master session key between itself and the UE. It also distributes application-specific keys to NAFs.
- **NAF** that is any kind of application server that is able to use GAA to authenticate subscribers. The NAF functionality is typically only a small part of the server software and can be, for example, realized through a NAF library. The NAF can be (i) part of the home operator network, (ii) in another operator network, or (iii) in a third-party network. We assume that in cases (ii) and (iii), the business relationship between the home operator and the owner of the NAF will be defined in a contractual agreement.
- **Terminal** or **User Equipment** (UE) that contains a smart card, e.g., UICC, issued by mobile network operator (MNO). The smart card contains also the long-term master key K that is shared with the operator.

These four elements form the core of GAA. Typically, there is one logical HSS and BSF per MNO but there are naturally multiple UEs and application servers, i.e., NAFs interworking in the system. If an HSS is used in a larger network with several HSS servers, then a Subscriber Locator Function (SLF) might be used. The BSF can query the SLF to determine the correct HSS for a given subscriber. However, when looking at a single GAA transaction, there is only one of each element involved at one time. The architecture of GAA with some details on components that are working inside each element is shown in Figure 3.5. Note that the SLF is not depicted in this figure.

UE: Terminals that are capable of using GAA need a GAA supporting function. One way to integrate the GAA supporting function is to add a dedicated software component to the operating system, which facilitates the communication with the

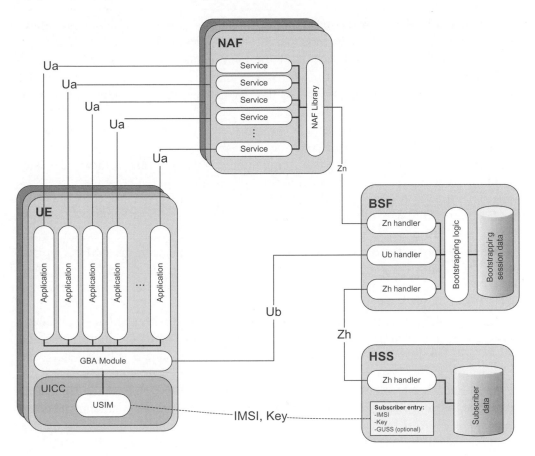

Figure 3.5.   GAA components

smart card (UICC) and the BSF on behalf of the applications utilizing GAA in the terminal. In our case, we call the GAA supporting function the 'GBA Module' (or also called 'BSF Client').

The GBA module is responsible for establishment of GAA master session key together with the BSF. This is carried out by performing the AKA procedure over the Ub reference point. The GBA module does its part by using the smart card inserted into the terminal. The smartcard is either a SIM card, or a UICC containing the USIM or ISIM applications.

The GBA module is responsible for communicating with the smart card via the device drivers, which are also part of the platform. It also offers an Application

Programming Interface (API) to the applications in the terminal through which these applications can request application-specific GAA keys from the GBA module so that they can be used with a NAF over the Ua reference point. It should be noted that the actual application using the GAA key may also reside outside the terminal and is connected locally, e.g., via USB cable or Bluetooth to the UE. This is then usually referred to as 'split terminal' case.

The high-level terminal architecture for GAA can be found in [TR33.905] and for Nokia Series60 devices in Section 5.2 of this book.

NAF: The NAF functionality (e.g., implemented by a software library) can easily be added to any server. The application server may reside in the network of the subscriber's home operator, but it can also be run by a third-party network. What the NAF functionality actually means is that a server is able to use GAA-based keys to authenticate users in Ua reference point and to support the key request on the Zn interface. In order to do so, it must be able communicate with the BSF to request these keys along with some other additional information which we come to later. The Zn interface for requesting the key can be based on the Diameter protocol [RCF3588] or Web Services [W3C-SOAP].

BSF: The BSF is either a dedicated server or combined with another set of nodes operated by an MNO. The terminal always bootstraps with his home credential server, i.e., the BSF that is owned by the MNO with whom the subscriber has the subscription.

HSS: Each MNO has some home server (for example, the HSS or the HLR) that stores the long-term keys of the subscriber. The only GAA specific data that the home server may store is the GUSS element. GUSS is a feature that can be used by any kind of service (but the support of the GUSS is mandatory only for a limited range of services). The BSF can request the GUSS and needed credential data from the HSS using the diameter-based Zh interface, if the HSS supports Diameter. If the HSS does not support the Diameter protocol, it may then use the Mobile Application Part (MAP)-based Zh′ interface.).

These are the nodes involved in the implementation of GAA. The GAA functionality can be split in two distinct phases: GAA bootstrapping and GAA authentication. First, the generation of the keys and then later on the usage of those. In more technical terms, we have:

- **GAA bootstrapping**, during which the UE, the BSF and the HSS work together to establish a shared secret (GAA master session key) utilizing reference points Ub between terminal and BSF and Zh between BSF and HSS (with potential detour using the SLF to find the right HSS).
- **GAA authentication**, during which the UE, a NAF and the BSF work together to mutually authenticate the subscriber using the application in the UE, and the service running a server with NAF functionality using reference point Ua between the terminal and the NAF, and Zn between the NAF and the BSF.

The GAA bootstrapping run is based on the cellular authentication that therefore serves as a prerequisite for the service-specific GAA authentication. One GAA bootstrapping run can be used for several GAA authentications, but before the service-specific GAA authentication can take place, there has to be at least one GAA bootstrapping run.

In GAA, the reference points Ub, Zh and Zn are specified in such a way that interoperability between different terminal manufacturers and BSF network node vendors can be assured, and the application server NAFs can interoperate with any BSF without any vendor-specific modifications. This allows application developers to connect their NAF server to different MNOs without being forced to change their application programs.

The Ua reference point has been more loosely specified. The intention is that any protocol could be a Ua reference point. If GAA is used in some form to authenticate and otherwise secure the connection between a client and a server, then this protocol is also a Ua reference point. Thus, the number of potential protocols that can implement the Ua reference point is huge. For example, if a protocol supports username/password-based authentication scheme, GAA can be plugged in and used there almost directly. This is the intention of GAA: to be able to use under lying authentication framework (e.g., (U)SIM cards) in different deployment scenarios and, in particular, make life easier for the users by not requiring to remember a large range of username/password combinations. Instead of remembering passwords, users are able to reuse their cellular authentication in a secure and automated manner. Therefore, the Ua reference point has not been selected to use one single protocol.

3GPP has specified how to use GAA in two general cases:

- HTTP Digest [RFC2617]; and
- Pre-Shared Key (PSK) TLS [RFC4279].

Already these two cover a wide range of use cases as HTTP [RFC2616] is the most widely used protocol in Internet today, and any other protocol can be tunnelled through TLS [RFC4346], which can run on top of both Transport Control Protocol (TCP) for normal TLS and User Datagram Protocol (UDP) for Datagram TLS (see IETF Datagram TLS specification [RFC4347]). These cases are covered in Section 4.1 of this book.

3.2.2 Bootstrapping

It is assumed that the terminal contacts an application server NAF at some point in time to register or request a service. The NAF server desires to secure the communication to the terminal using GAA. The bootstrapping phase is not dependent on the actual service to be used. The NAF indicates the desire to use GAA to the NAF application in the terminal (e.g., web browser or video streaming application). Then the GAA bootstrapping is triggered by the application in the terminal (or alternatively the GBA module in the terminal can be configured to have a valid GAA master session key at all time, i.e., 'keep-alive GAA session'). The GBA module will perform the GAA bootstrapping over Ub interface if it does not have a valid GAA master session key Ks when an application requests application-specific GAA key from it.

The GAA master session key Ks is identified by a key identifier B-TID. It is valid for a certain amount of time after it has been established between the GBA module in the UE and the BSF. The lifetime of the key is set by an operator and may depend on factors like load balancing, value of the provided services and trustworthiness of the application servers connected to the BSF. Section 5.4.3 discusses the effect key lifetime setting has on the amount of bootstrapping operations.

Below is the message sequence flow diagram in the case where the UE contains a 3G smart card, i.e., a UICC, that does not have any GAA specific functionality and the subscriber database is the HSS. The smart card hosts either a 3G USIM or ISIM application. This configuration (or GBA variant) is also referred as GBA_ME or normal GBA where ME refers to mobile equipment and indicates that all GAA functionality, e.g., key generation and storage functionality resides in a trusted part of the ME.

It should be noted that the GBA module will always establish the GAA master session key with the home BSF, i.e., it will contact the BSF that is operated by the same operator that issued the UICC to the subscriber. The address of the home BSF is derived the following way:

- If USIM or SIM application on the UICC is used, the application identifier is IMSI. If the IMSI is, for example, 234150999999999, where the Mobile Country Code (MCC) is 234, and the Mobile Network Code (MNC) is 15, and the Mobile Station Identity Number (MSIN) is 0999999999, then the BSF address is derived to be bsf.mnc015.mcc234.pub.3gppnetwork.org.
- If ISIM application on the UICC is used, the subscriber identifier is IMPI. If the IMPI is subcriber@operatorB.com for example, then the derived BSF address is bsf.operatorB.com.

The derivation of BSF address from IMSI or IMPI is specified in [TS23.003]. (See Figure 3.6.)

The general form of the KDF used in GAA is:

derived key = KDF (*master session key*, *gba variant*, RAND, IMPI, NAF_ID).

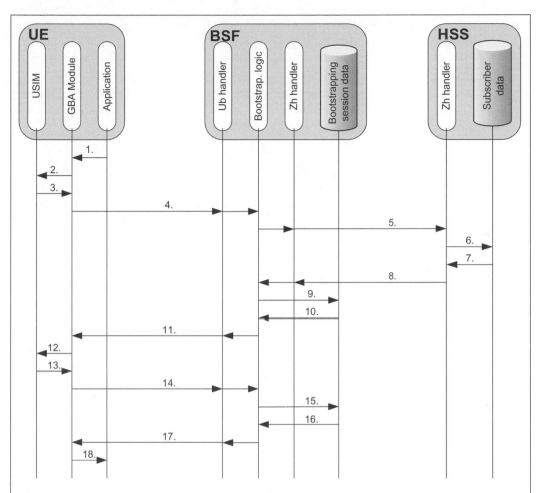

Figure 3.6.   Normal bootstrapping message flow

1. An application in the UE has come to a state that it needs application-specific GAA keys from the GBA module in the UE. Thus, it contacts the GBA Module in the UE and gives the NAF_ID parameter to it. The NAF_ID consists of the Fully Qualified Domain Name (FQDN) of the NAF application server (i.e., DNS name of the server) and a five-octet Ua security protocol identifier that identifies the protocol that is being used between the UE and the NAF over the Ua reference point.

2. The GBA module contacts the USIM application in the UICC and asks for the International Mobile Subscriber Identity (IMSI) of the USIM. IMSI is the unique identifier of the USIM application that is used in HSS to locate the corresponding long-term key of USIM, i.e., to find the counterpart key of the subscriber in the subscriber database. If an ISIM application is used, then the unique identifier is called IP Multimedia Private Identity (IMPI).

3. The USIM returns the IMSI to the GBA module. The GBA module converts the IMSI to the IMPI format according to [TS23.003] so that it can be carried over the Ub

Figure 3.6. Normal bootstrapping message flow (*continued*)

reference point. In ISIM case, no conversion is needed because the identity is already in IMPI format.

4. The GBA module starts the bootstrapping over Ub reference point with the BSF by sending the IMPI to it. The message complies with the HTTP Digest AKA Version 1 messaging according to [RFC3310]: It should be noted that it is assumed that the password in 'AKAv1' HTTP Digest AKA is in binary format.

```
GET / HTTP/1.1
Host: bsf.home1.net:80
User-Agent: Bootstrapping Client Agent; Release-6
Date: Thu, 25 Oct 2007 2:38:00 GMT
Accept: */*
Authorization: Digest
        username="<IMPI>",
        realm="bsf.home1.net",
        nonce="",
        uri="/",
        response=""
```

5. The BSF receives the IMPI, converts it back to IMSI format if needed and sends a request for Authentication Vector (AV) and GUSS to the HSS over the operator internal diameter-based Zh interface. The GUSS request is optional, but if the operator wants also to use GBA_U variant of GAA, then the GUSS is required for the BSF to process the AV correctly.

6–7. The HSS queries its internal database for the master session key, next sequence number and GUSS data element. It then generates a random number, and with it, master session key, and the sequence number generates the AV which consists of RAND, AUTN (containing sequence number and a MAC), RES, CK and IK.

8. The five values of the AV (quintet) and, if available, the GUSS, are sent back to the BSF.

9–10. The BSF initializes the generation and storage of the bootstrapping session data to its internal session database. The actual storage and data management in the BSF is vendor implementation specific, but at least the AV, GUSS and key generation timestamp should be stored for further requests and processing.

11. The BSF challenges the UE by sending the RAND and AUTN to the UE to perform the client authentication:

```
HTTP/1.1 401 Unauthorized
Server: Bootstrapping Server; Release-6
Date: Thu, 25 Oct 2007 2:38:00 GMT
WWW-Authenticate: Digest
        realm="bsf.home1.net",
        nonce="base64(RAND + AUTN + serverdata)",
```

Figure 3.6. Normal bootstrapping message flow (*continued*)

```
        algorithm=AKAv1-MD5,
        qop="auth-int",
        opaque="5ccc...0c41"
```

12. The GBA module extracts the RAND and AUTN from the challenge and runs
    AUTHENTICATE command with the USIM with RAND and AUTN as input
    parameters to the smart card command [TS31.102] (in case a ISIM is used, then
    [TS31.103] applies).
13. Upon successful validation of RAND and AUTN, the USIM returns RES, CK and
    IK to the GBA module. The GBA module establishes the GAA master session key
    (Ks) by concatenating CK and IK received from the card.
14. The GBA module sends response back to the BSF by calculating the response using
    RES as password.

```
GET / HTTP/1.1
Host: bsf.home1.net:80
User-Agent: Bootstrapping Client Agent; Release-6
Date: Thu, 25 Oct 2007 2:38:00 GMT
Accept: */*
Authorization: Digest
        username="<IMPI>",
        realm="bsf.home1.net",
        nonce="base64(RAND + AUTN + serverdata)",
        uri="/",
        qop=auth-int,
        nc=00000001,
        cnonce="8802...308c",
        response="6629...4cf1",
        opaque="5ccc...0c41",
        algorithm=AKAv1-MD5
```

The BSF validates the response received from the UE with the stored data.

15–16. Upon successful validation, the BSF updates the bootstrapping session data by setting
       up key lifetime and generating the key identifier B-TID.
   17. The BSF sends B-TID and key lifetime to the UE using the HTTP payload, which
       is protected by the digest operations, as the quality of protection (qop) parameter is
       set to 'auth-int':

```
HTTP/1.1 200 OK
Server: Bootstrapping Server; Release-6
Authentication-Info: qop=auth-int,
        rspauth="23ac...5ff0",
        cnonce="8802...308c",
        nc=00000001,
```

Figure 3.6.   Normal bootstrapping message flow (*continued*)

```
              opaque="5ccc...0c41",
              nonce="base64(RAND + AUTN + serverdata)"
Date: Thu, 25 Oct 2007 2:38:00 GMT
Expires: Thu, 25 Oct 2007 14:38:00 GMT
Content-Type: application/vnd.3gpp.bsf+xml
Content-Length: 185

<?xml version="1.0" encoding="UTF-8"?>
<BootstrappingInfo xmlns="uri:3gpp-gba">
  <btid>base64(RAND)@bsf.operator.com</btid>
  <lifetime>2007-10-25T14:38:00Z</lifetime>
</BootstrappingInfo>
```

Upon receiving the message, the GBA module validates the response and stores the master session key Ks, the identifier B-TID and the corresponding key lifetime.

18. The GBA module derives the application-specific GAA key using the NAF_ID that was received from the application in step 1. It then returns this key with key lifetime, and optionally, other parameters, such as GAA bootstrapping type, which in this case would be 'GBA_ME'. In GBA_ME, only one application-specific GAA key is generated (Ks_NAF).

The Ks_NAF is computed as:

$$Ks\_NAF = KDF (Ks, 'gba\text{-}me', RAND, IMPI, NAF\_ID)$$

The KDF consists of an application of HMAC-SHA-256 and its exact form is defined in Annex B of [TS33.220]. The *master session key* is always Ks when deriving application-specific GAA keys. The *gba variant* parameter specified which GBA variant is in question. In GBA_ME case, its value is 'gba-me'. The other parameters consist of user's IMPI, the NAF_ID and RAND. The NAF_ID is constructed as follows:

$$NAF\_ID = FQDN \text{ of the } NAF \parallel Ua \text{ security protocol identifier.}$$

The Ua security protocol identifier specifies what kind of protocol is used in the Ua interface and the standardized value are specified in Annex H of [TS33.220]. The Ua security protocol identifier consists of five bytes, and for example, if HTTPS is used as the Ua protocol, that value would be (0x01, 0x00, 0x00, 0x00, 0x02) and if MBMS is used then it would be (0x01, 0x00, 0x00, 0x00, 0x01) for MBMS. The main reason for using the Ua security protocol identifier is to have key diversification between different protocols if there are multiple protocols used between the UE and the NAF. This way, if due to some weakness in one protocol the Ks_NAF is discovered, it would affect another protocol as its Ks_NAF would be different.

Other variants of GAA bootstrappings are described in Section 3.3. The difference between these variants results from the underlying security architecture.

3.2.3 Authentication

In this phase, the key generated in the bootstrapping run are actually used for authentication. As said earlier, the protocol used in Ua reference point, i.e., between the UE and the BSF, can be any kind of protocol. Below is a generic message sequence flow on how application-specific GAA key (Ks_NAF) is used for authentication purposes. (See Figure 3.7.)

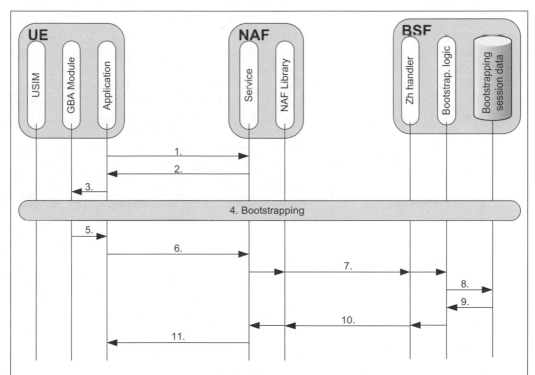

Figure 3.7. Authentication Message Flow Using GAA

  1. The application in the UE contacts an NAF using an application-specific protocol over the Ua interface. It may indicate that it can use GAA although this is not required.
  2. The NAF decides to use GAA for authentication based on its internal policy or upon a requirement in the service agreement with the operator. This decision may depend

Figure 3.7.   Authentication Message Flow Using GAA (*continued*)

on whether GAA support was indicated during step 1, or it can be static rule to always use GAA. The NAF then sends a request with required parameters to challenge the UE to authenticate using GAA.

3. This step is the same as step 1 in previous section. The application in the UE requests Ks_NAF from the GBA module and send the application server identifier the NAF_ID to the GBA module. The interface between the GBA module and the application in the UE is secured through platform security, more on this subject can be found in Chapter 5.

4. If GBA module does not have a valid bootstrapping session active, then it will perform GAA bootstrapping run at this phase as outlined in the previous section (starting directly from step 2 because step 1 is already done). If it already has the session, then it can proceed directly to step 5 without making a bootstrapping run.

5. This is the same step as step 18 in previous section. The GBA module returns the derived Ks_NAF, the key identifier, i.e., B-TID, and the key lifetime to the application in the UE (e.g., browser or streaming application) through the GBA module API.

6. The application in the UE uses the key to calculate a response to the challenge of the NAF. It will then include this response with the B-TID to the message and send it to the NAF application server.

7. The NAF will extract the B-TID, response, and any other necessary information from the message. It will then request the Ks_NAF from the BSF over the Web Services (WS) or diameter-based Zn interface. It will include at least B-TID and NAF_ID to the request, but may also request one or more User Security Settings (USS) elements from the BSF, if it is configured to do so. GUSS and USS usage is discussed in Section 3.3. The USS may also contain information about whether 2G GBA has been used for bootstrapping or not (see next section for description of 2G GBA), the NAF may then compare this information with its local security policy.

8–9. The BSF retrieves the bootstrapping session information from its local session database including the session key Ks, key lifetime and optional USS elements if requested from the NAF.

10. The BSF derives the Ks_NAF based on Ks, NAF_ID and other necessary parameters. It then returns this key, key lifetime and USS elements (if requested and existent in the HSS/BSF) to the NAF.

11. The NAF validates the received response from the UE using the Ks_NAF, and if the validation is successful, it creates an authenticated session for the user in its local database and indicates to the user that the authentication was successful.

Naturally, the NAF and the application in the UE need to run many application-specific functionalities and tasks, but in essence, the above is what they need to do from GAA's perspective. An example of an additional application-specific task could be authentication of the NAF by the UE using the same GAA key.

More application-specific message sequence flows for several use cases are described in Chapter 4.

3.3 Variations of the Generic Bootstrapping Architecture

GAA builds upon the shared key stored in MNO's database and in the smart card that is given out to the user. Today's smart cards have a much broader functionality than the SIM cards that users got a few years ago. The second generation Global System for Mobile Communication (GSM) SIM card specification was frozen by 3GPP in Release 4 in 2001 [TS51.011], i.e., only essential corrections[1] are now allowed. The third generation smart card is the Universal Integrated Circuit Card (UICC), which was modelled after the GSM SIM card, but can host several applications. It contains the USIM [TS31.102] that is used for authentication and is the evolution of GSM SIM.

For 3GPP2 systems that are, for example, used in North America, the UICC can contain, in addition to (or instead of) the USIM application, the Removable User Identity Module (R-UIM and ISIM) application. The UICC card can also run the GSM SIM card functionality or the IP Multimedia Subsystem (IMS) Identity Module (ISIM) [TS31.103] as an application.

If every mobile terminal would include a smart card with USIM application, a single variant of GAA could work everywhere. But large-scale replacement of deployed smart cards with new model is costly and so smart cards that were given out to the subscribers are rarely replaced. This is one of the reasons why several bootstrapping variants have been standardized.

The bootstrapping procedure that we have described so far was standardized first. It is now called GBA_ME or 'normal GBA'. GBA_ME bootstraps from USIM (UMTS) credentials and the bootstrapped keys are stored in the mobile equipment (ME), outside the smart card.

The second bootstrapping variant is called GBA_U or 'UICC-aware' bootstrapping. It was created because for some services, e.g., broadcast mobile TV, the bootstrapped key should be protected not only from outsiders but also from the owner of the mobile device. For that reason, the bootstrapped master key stays on the smart card; it is not revealed to applications on the ME.[2]

---

[1] An essential correction is a correction without which the system would not work.
[2] As we will describe in detail later, in GBA_U, the smart card creates two keys for each application: external and internal. The external key is given to application and can be used in the same way as in GBA_ME. The internal key stays in the smart card. The application that needs that key, e.g., to decrypt data, will send the encrypted data to the smart card.

The other GBA variants were created to enable bootstrapping in second generation, i.e., pre-UMTS, cellular networks.

The third GBA variant is called '2G GBA' or 'SIM GBA' and bootstraps from SIM (GSM) credentials.

The fourth, the fifth and the sixth GBA variants were standardized by 3GPP2 for Code Division Multiple Access (CDMA) networks.

3.3.1 GBA_ME

This is the terminal-based variant of GAA, also sometimes called normal GBA and was outlined in full detail in the previous section. In GBA_ME, it is assumed that the terminal contains a 3G UICC smart card, but the card does not need to be aware of GAA or to be specially configured for that purpose. The UICC contains an ISIM, or USIM or R-UIM application; in most typical scenarios, it will contain at least a USIM. The cryptographic key management of GAA is done by the terminal, in particular, the master session key and the application-specific keys are derived in the terminal, after requesting the needed key derivation material from the UICC [TS33.220]. In GBA_ME, only one application-specific key Ks_NAF per NAF_ID is derived from the master session key Ks that has been established in each bootstrapping run with the BSF. The NAF_ID consists of the FQDN name of the NAF and of the Ua security protocol identifier used between the NAF and the terminal. The security of the master key and the application-specific key are protected by the terminal platform and the operating system. The interface between the UICC and the terminal is described in [TS31.102] for the USIM and in [TS31.103] for the ISIM. The message flow and the details were explained in the Section 3.2.

3.3.2 GBA_U

GBA_U is the smart card centred variant of GAA. It requires a specially configured UICC smart card that is GAA aware and supports some GAA-specific interfaces and functionalities. The difference compared to GBA_ME is that the master key Ks stays inside the UICC and all key derivations happen in the UICC. If GBA_U is used, then the BSF and the smart card derive two keys, the Ks_int_NAF and the Ks_ext_NAF, instead of just a single key Ks_NAF. The Ks_int_NAF stays in the card and the Ks_ext_NAF is given out to the terminal. GBA_U is, for example, used for MBMS security, as an alternative to GBA_ME. The key derivation varies slightly from the one for GBA_ME:

$$Ks\_ext\_NAF = KDF (Ks, \text{'gba-me'}, RAND, IMPI, NAF\_ID)$$

and

$$Ks\_int\_NAF = KDF (Ks, \text{'gba-u'}, RAND, IMPI, NAF\_ID).$$

The KDF is the HMAC-SHA-256-based Key Derivation Function and the NAF_ID is a concatenation of the FQDN of the NAF and the Ua security protocol identifier (see Section 3.2).

Below is message sequence flow diagram in the case where the UE contains a UICC that contains an USIM application and supports GAA-specific functionality (i.e., the smart card is GBA_U enabled) and the subscriber database is an HSS. (See Figure 3.8.)

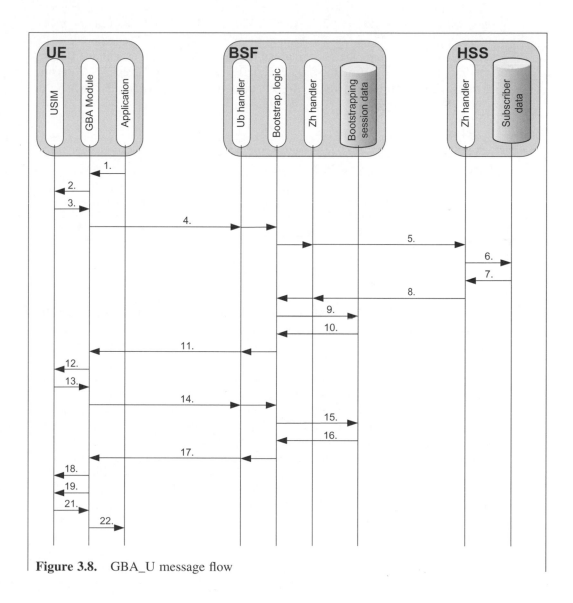

Figure 3.8.   GBA_U message flow

Figure 3.8. GBA_U message flow (*continued*)

1. An application in the UE has come to a state that it needs application-specific GAA keys from the GBA module in the UE. Thus, it contacts the GBA module in the UE and gives it the NAF_ID. The NAF_ID consists of the FQDN of the NAF application server and a Ua security protocol identifier.
2. The GBA module contacts the USIM application in the UICC and asks for the International Mobile Subscriber Identity (IMSI) of the USIM. IMSI is the unique identifier of the USIM application that is used in HSS to find the counterpart key of the subscriber in the subscriber database.
3. The USIM returns the IMSI to the GBA module. The GBA module converts the IMSI to the IMPI format so that it can be carried over the Ub reference point.
4. The GBA module starts the bootstrapping over Ub reference point with the BSF by sending the IMPI to it. The message complies with the HTTP Digest AKA Version 1 messaging according to [RFC3310]. An example message is given below:

```
GET / HTTP/1.1
Host: bsf.home1.net:80
User-Agent: Bootstrapping Client Agent; Release-6
Date: Thu, 25 Oct 2007 2:38:00 GMT
Accept: */*
Authorization: Digest
        username="<IMPI>",
        realm="bsf.home1.net",
        nonce="",
        uri="/",
        response=""
```

5. The BSF receives the IMPI, converts it back to IMSI format, and sends a request for Authentication Vector (AV) and GBA User Security Settings (GUSS) to the HSS over the diameter-based Zh interface. Since GBA_U is deployed in the operator network, the HSS needs to return the GUSS.
6–7. The HSS queries its internal database for the master session key, next sequence number, and GUSS data element. It then generates a random number, and with it, master session key, and the sequence number generates the AV, which consists of RAND (the random number), AUTN (sequence number, and a MAC), RES (the expected response value), CK (cipher key) and IK (integrity key).
8. The five values of the AV (quintet) and, if available, the GUSS is sent back to the BSF. The GUSS contains information that in the XRES the least significant bit has to be flipped before storing it and that the MAC* and AUTN* have to be calculated as outlined in Chapter 5 of [TS33.220]. It should be noted that without the GUSS, the BSF would have no indication that it has to do these changes, i.e., for GBA_U support the GUSS is needed in the BSF.

Figure 3.8. GBA_U message flow (*continued*)

The flipping of the least significant bit of XRES value above is done because then the corresponding AV cannot be used in the normal AKA procedure. With this change, it can be ensured that this particular AV can be used only in the GBA_U mode.

9–10. The BSF initializes the generation and storage of the bootstrapping session data to its internal session database. The actual storage and data management in the BSF is vendor implementation-specific, but at least the AV, GUSS and key generation timestamp should be stored for further requests and processing.

11. The BSF challenges the UE by sending the RAND and AUTN* to the UE to perform the client authentication:

```
HTTP/1.1 401 Unauthorized
Server: Bootstrapping Server; Release-6
Date: Thu, 25 Oct 2007 2:38:00 GMT
WWW-Authenticate: Digest
        realm="bsf.home1.net",
        nonce="base64(RAND + AUTN* + serverdata)",
        algorithm=AKAv1-MD5,
        qop="auth int",
        opaque="5ccc069c403ebaf9f0171e9517f30e41"
```

12. The GBA module extracts the RAND and AUTN* from the challenge, and runs AUTHENTICATE command with the UICC in 'GBA Derivation' mode with RAND and AUTN* as input parameters to the smart card command [TS 31.102]. In the 'GBA Derivation' mode, the UICC does not return the CK and IK back to me opposite to the UMTS Mode where they are returned. The UMTS Mode is used with GBA_ME, but the 'GBA Derivation' mode is used with GBA_U.

The UICC validates the RAND and calculates IK and MAC by performing:

$$MAC = MAC^* \oplus Trunc (SHA\text{-}1 (IK)).$$

Then the UICC checks AUTN value (i.e., $SQN \oplus AK \parallel AMF \parallel MAC$) to check that the challenge was really coming from an authorized network.

13. Then the UICC calculates CK and RES. This will result in the master session key Ks, which is a concatenation of CK and IK that is then stored in the UICC. The UICC then transfers only the RES (after flipping the least significant bit) to the GBA module.

14. The GBA module sends response back to the BSF by calculating the response using RES as password.

```
GET / HTTP/1.1
Host: bsf.home1.net:80
User-Agent: Bootstrapping Client Agent; Release-6
Date: Thu, 25 Oct 2007 2:38:00 GMT
Accept: */*
```

Figure 3.8. GBA_U message flow (*continued*)

```
Authorization: Digest
    username="<IMPI>",
    realm="bsf.home1.net",
    nonce="base64(RAND + AUTN* + server specific data)",
    uri="/",
    qop=auth-int,
    nc=00000001,
    cnonce="6629fae49393a05397450978507c4ef1",
    response="6629fae49393a05397450978507c4ef1",
    opaque="5ccc069c403ebaf9f0171e9517f30e41",
    algorithm=AKAv1-MD5
```

The BSF validates the response received from the UE with the stored data.

15–16. Upon successful validation, the BSF updates the bootstrapping session data by setting up key lifetime and generating the key identifier B-TID.

17. The BSF sends B-TID and key lifetime to the UE using the HTTP payload.

```
HTTP/1.1 200 OK
Server: Bootstrapping Server; Release-6
Authentication-Info: qop=auth-int,
        rspauth="6629fae49394a05397450978507c4ef1",
        cnonce="6629fae49393a05397450978507c4ef1",
        nc=00000001,
        opaque="5ccc069c403ebaf9f0171e9517f30e41",
        nonce="base64(RAND + AUTN* + serverdata)",
        qop=auth-int
Date: Thu, 25 Oct 2007 2:38:00 GMT
Expires: Thu, 25 Oct 2007 14:38:00 GMT
Content-Type: application/vnd.3gpp.bsf+xml
Content-Length: 185

<?xml version="1.0" encoding="UTF-8"?>
<BootstrappingInfo xmlns="uri:3gpp-gba">
   <btid>user@bsf.operator.com</btid>
   <lifetime>2007-10-25T14:38:00Z</lifetime>
</BootstrappingInfo>
```

Upon receiving the message, the GBA module validates the response, i.e., checks the 'rspauth' value, and stores the NAF-specific Ks_ext_NAF key, the identifier B-TID and the corresponding key lifetime.

18. To conclude the bootstrapping phase of the procedure, the GBA module stores the B-TID, the RAND, and the key lifetime received in the previous step to the UICC.

Figure 3.8. GBA_U message flow (*continued*)

19. To respond to the original request made by the application in step 1, the GBA module derives application-specific GAA keys. This key derivation happens in the UICC. The GBA module sends the NAF_ID (received in step 1), and the IMPI to the UICC, and requests it to derive the keys Ks_ext_NAF and Ks_int_NAF.
20. The UICC uses the GAA master session key Ks and other parameters as outlined in [TS33220] to derive both the Ks_int_NAF and the Ks_ext_NAF. The UICC stores the Ks_int_NAF together with the corresponding B-TID and RAND on its file system. It then returns the Ks_ext_NAF to the GBA module.
21. The GBA module sends now the Ks_ext_NAF key with key lifetime and optionally other parameters, such as GAA bootstrapping type, which in this case would be 'GBA_U' to the application in the UE. If the application wants to utilize later the Ks_int_NAF, then it has only the possibility to hand over encrypted data to the smart card for processing with Ks_int_NAF since the Ks_int_NAF key does not leave the card.

3.3.3 2G GBA

Though smart cards are evolving and have new features with each release, cards that have been given to subscribers are rarely changed. Thus, most mobile operators have a large number of subscribers that have use SIM card (implemented according to [TS51.011]) and operators want to exploit these cards and the existing infrastructure also for new services. Therefore, 3GPP decided that service consumption of GAA-secured services should also be possible with a SIM card and later on integrated also the case that the operator may use an HLR (for more information see [TS33.220]). The main requirement was that no changes to existing cards need to be made and that users could just use their old cards in new terminals to consume a service (e.g., watch mobile TV). This 2G GBA, also called SIM GBA, was then specified in Annex I in [TS33.220] in Release 7 of 3GPP. Additionally, [TR33.920] contains a list of changes that need to be implemented for 2G GBA on top of Release 6 [TS33.220].

The core network needs to have an HSS or an HLR. The BSF needs to retrieve the 2G Authentication Vector and optionally the GUSS from the HSS using a diameter-based interface. If an HLR is deployed instead of the HSS or an HSS without Diameter support is used, then only the 2G Authentication Vector is retrieved via a MAP-based interface. 2G GBA is very similar to GBA_ME; in 2G GBA also, the master key and the application-specific key are derived and stored in the terminal and protected by the platform and the operating system.

The Key Derivation Function (KDF) is the same as for GBA_ME, but the input parameter 'key' is different from the GBA_ME case due to the different format of the baseline key Kc. The generated application-specific key Ks_NAF has the same format as the GBA_ME-based key and can be used in the same way. Terminals that support 2G GBA are also recommended to support GBA_ME and GBA_U. When the UICC card supports both the SIM application functionality and the 3G application functionality (USIM or ISIM), the terminal is required to use the 3G application.

The application in the terminal requesting the NAF-specific key can indicate to the GBA module in the platform whether 2G GBA is acceptable for this application or not. In general, it is assumed that 2G GBA can be used for applications, except if is explicitly prohibited by the operator or by the application specification. It is expected that the application server in the network (i.e., the NAF server) has a local policy about whether to accept 2G subscribers, but this depends on the actual security needs of the application which uses the keys.

It should be noted that in case where a terminal can use either a 2G SIM-based GBA or a 3G USIM-based GBA, the UE always has to choose use of the USIM for the bootstrapping. Both 2G GBA and the MAP-based Zh' interface between BSF and HLR are 3GPP Release 7 features and are optional to support. We will now outline the message flow for 2G GBA that is used together with an HLR. (See Figure 3.9.)

It should be noted that even if the BSF and the GBA module store the key, a situation may occur that either the BSF or the UE has deleted the key and the related material. In the BSF, this might be caused by memory management and the terminal may have deleted the data, e.g., when UICC was removed. If the BSF has deleted the key, then NAF will get the 'B-TID unknown' error reply after requesting the key using the B-TID over Zn reference point. In this case, the NAF will indicate to the UE that it needs to perform new GAA bootstrapping, to derive new Ks_NAF from newly generated Ks, and to request Ks_NAF from the BSF again. If the UE has deleted the key, it will perform new GAA bootstrapping when a Ks_NAF is needed. This will cause the BSF to overwrite the existing key in its databases as the BSF is required to remember only the most recent key and the related material for a particular subscriber.

3.3.4 Detection of Bootstrapping Variants by the NAF

The application server (NAF) may not be part of operator's network, hence, it is not directly aware of the GBA variant used between the BSF and the UE, and the resulting security level of the keys. It is not that straightforward for the NAF to determine the provided security level. There are several cases to be distinguished. In the following, we show how the NAF can determine which case is in use, i.e., what is the security level provided:

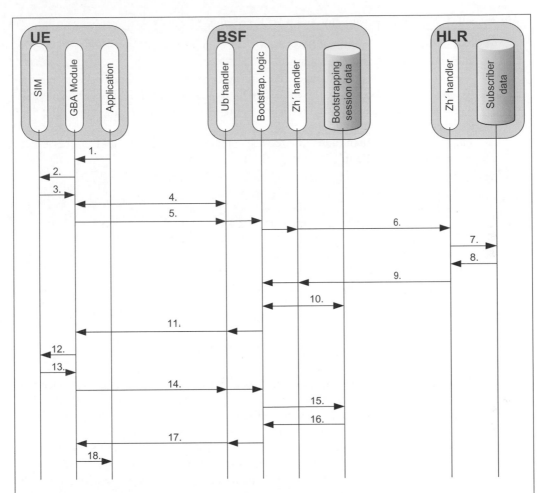

Figure 3.9. 2G GBA message flow with HLR

The message flow for this scenario is similar to the one described for normal GAA in the previous section, but has some particularities.

1. An application in the UE has come to a state that it needs application-specific GAA keys from the GBA module in the UE. Thus, it contacts the GBA module in the UE and gives it the NAF_ID.
2. The GBA module contacts the SIM card and asks for the IMSI.
3. The SIM returns the IMSI to the GBA module. The GBA module converts the IMSI to the IMPI format so that it can be carried over the Ub reference point.
4. The UE sets up a server-authenticated TLS tunnel to the BSF on the Ub interface. The authentication of the BSF is performed using certificates. The further communication between ME and BSF is sent through this secure TLS tunnel.

   The TLS usage is the first difference between 2G GBA and GBA_ME. TLS is required because the BSF needs to be authenticated using a certificate that has been

Figure 3.9.   2G GBA message flow with HLR (*continued*)

issued by a trusted certification authority (CA). In GBA_ME, the BSF is indirectly authentication as part of the AKA.

Also upon receiving messages, both the UE and the BSF should check that the 'realm' parameter contains the same FQDN that is present in BSF's certificate that was used to setup the TLS tunnel. The 'realm' check is performed to make sure that there is no man-in-the-middle.

5. The UE now sends an initial HTTP request containing subscriber's IMPI. An example request may look like this:

```
GET / HTTP/1.1
Host: bsf.home1.net:80
User-Agent: Bootstrapping Client Agent; Release-7
Date: Thu, 25 Oct 2007 2:38:00 GMT
Accept: */*
Authorization: Digest
        username="<IMPI>",
        realm="bsf.home1.net",
        nonce="",
        uri="/",
        response=""
```

6. The BSF converts IMPI back to IMSI format. Then the BSF requests the authentication vector from the HLR using the Zh' reference point as outlined in [TS29.109]. No GUSS or USS is requested since the HLR is not designed to store the GUSS. If an operator desires, he may have an additional database connected to the BSF that serves as a GUSS repository, but this database and the discovery of the GUSS for this case are not covered by the standard. If an operator has a mixed infrastructure and several HLR and HSS deployed in his network, then the BSF may need to determine which protocol to use for which subscriber. This selection process is also not covered by the standard, but left for operator-specific implementations and configuration, e.g., look-up server or enhanced SLF server.

7–8. The HLR retrieves the authentication vector from its internal database.

9. Since we consider here the SIM-based GBA variant, the HLR returns the requested 2G authentication vector AV = (RAND, SRES, Kc) over the Zh' reference point. Latest at this point of time, the BSF knows that the UE is equipped with a SIM card by looking at the type of the received authentication vector. It should be noted that the MAP protocol supports 3G and 2G authentication vector requests. The Zh' reference point is used when the HSS does not support the diameter-based Zh or when there is only an HLR available.

10. The BSF converts the received 2G authentication vector (AV = (RAND, SRES, Kc) to the 2G GBA-specific RES by setting

$$RES = KDF \ (key, \ \text{'3gpp-gba-res'}, \ SRES),$$

which is truncated to 128 most significant bits, and where key is

Figure 3.9.   2G GBA message flow with HLR (*continued*)

key = Kc || Kc || RAND.

Also the BSF selects a 128-bit random number 'Ks-input' and sets the server-specific data equal to the Ks-input in the AKA-nonce of HTTP Digest AKA and stores the data

11. The BSF sends the RAND, the AUTN (which contains only 0×00 octets), and the Ks-input as the server-specific data through the TLS tunnel in the 401 message. This in turn triggers the UE to authenticate itself.

```
HTTP/1.1 401 Unauthorized
Server: Bootstrapping Server; Release-7
Date: Thu, 25 Oct 2007 2:38:00 GMT
WWW-Authenticate: Digest
      realm="bsf.home1.net",
      nonce="base64(RAND + AUTN(=0x00..00) + Ks-input)",
      algorithm=AKAv1-MD5,
      qop="auth-int",
      opaque="5ccc069c403ebaf9f0171e9517f30e41"
```

12–13. The GBA module in the UE obtains the RAND from the received message and then communicates with the SIM card to calculate the corresponding Kc, SRES and also the 2G GBA-specific RES.

14. The GBA module in the UE sends the HTTP Digest AKA response message to the BSF using the GBA-specific RES value as a password.

```
GET / HTTP/1.1
Host: bsf.home1.net:80
User-Agent: Bootstrapping Client Agent; Release-7
Date: Thu, 25 Oct 2007 2:38:00 GMT
Accept: */*
Authorization: Digest
      username="<IMPI>",
      realm="bsf.home1.net",
      nonce="base64(RAND + AUTN(=0x00..00) + Ks-input)",
      uri="/",
      qop=auth-int,
      nc=00000001,
      cnonce="6629fae49393a05397450978507c4ef1",
      response="6629fae49393a05397450978507c4ef1",
      opaque="5ccc069c403ebaf9f0171e9517f30e41",
      algorithm=AKAv1-MD5
```

15. The BSF can now authenticate the UE by verifying the HTTP Digest AKA response message received. Then the BSF generates the master session key Ks:

Figure 3.9.   2G GBA message flow with HLR (*continued*)

$$Ks = KDF \text{ (key, Ks-input, '3gpp-gba-ks', SRES).}$$

The BSF also generates the key identifier B-TID value from the base64 encoded RAND value and the BSF server name, i.e., the B-TID is of the form

B-TID = base64encoded(RAND)@BSF_servers_domain_name.

The BSF stores the GAA session master key, the B-TID and the associated key lifetime.

16. The BSF returns the B-TID and the key lifetime of the GAA session master key and indicates the success of the authentication.

```
HTTP/1.1 200 OK
Server: Bootstrapping Server; Release-7
Authentication-Info: qop=auth-int,
        rspauth="6629fae49394a05397450978507c4ef1",
        cnonce="6629fae49393a05397450978507c4ef1",
        nc=00000001,
        opaque="5ccc069c403ebaf9f0171e9517f30e41",
        nonce="base64(RAND + AUTN(=0x00..00) + Ks-input)",
        qop=auth-int
Date: Thu, 25 Oct 2007 2:38:00 GMT
Expires: Thu, 25 Oct 2007 14:38:00 GMT
Content-Type: application/vnd.3gpp.bsf+xml
Content-Length: 185

<?xml version="1.0" encoding="UTF-8"?>
<BootstrappingInfo xmlns="uri:3gpp-gba">
   <btid>user@bsf.operator.com</btid>
   <lifetime>2007-10-25T14:38:00Z</lifetime>
</BootstrappingInfo>
```

17. Upon receiving the message, the GBA module validates message by checking the 'rspauth' value. If this validation is successful, the GBA module generates the master session key Ks in the same manner as the BSF did in step 15. The GBA module stores then the GAA master session key, the key lifetime and the B-TID.

18. The GBA module derives from the GAA master session key Ks, the application-specific key Ks_NAF and sends it to the application so that the application can use the key to secure the communication over Ua reference point.

(a) The terminal contains a SIM card and uses 2G GBA:

If the NAF is able to use GBA_U, it would send the GBA_U awareness indicator over the Zn reference point to the BSF. Since the BSF uses 2G GBA, it would, in this case, return in the ME Key material field the Ks_NAF. The UICC key material field would not be sent to the NAF. Additionally, the BSF can utilize the User Security Settings (USS) to indicate which key to use by populating the KeyChoice field in the USS stating if 'Ks_NAF or Ks_ext_NAF' is supposed to be used. The NAF receives also indication that the SIM has been used by setting the GBA Type equal to 1 [TS29.109].

If the NAF is not GBA_U-aware and sends no indicator to the BSF or sets the indicator accordingly, then the BSF returns in the ME key material field the Ks_NAF and the UICC-Key-Material field would not be sent. The BSF may send the USS information if saying 'Ks_NAF or Ks_ext_NAF' is supposed to be used (not the population of the actual fields does not have these values). As in the previous case, the NAF will also get indication that SIM has been used since the GBA Type would be set to 1. The GBA Type field is only for indication if SIM-based GBA has been used or not.

(b) The terminal contains a GBA_U unaware UICC and uses GBA_ME:

If the NAF is able to use GBA_U keys, then it would send a GBA_U awareness indication to the BSF. The BSF returns then in the ME key material field the GBA_ME key Ks_NAF since the card in the terminal is not GBA_U aware. The UICC key material field would not be populated by the BSF. The BSF may additionally send the USS, containing the key usage field and the GBA-type field.

If the NAF is not GBA_U-aware, it sends no indicator to the BSF or sets the indicator accordingly. In this case, the BSF returns in the ME key material field the GBA_ME key Ks_NAF and the UICC key material field would be empty. The BSF may send the USS information stating if 'Ks_NAF or Ks_ext_NAF' are supposed to be used and that no SIM has been used (GBA Type = 0).

(c) The terminal contains a GBA_U-aware UICC and uses GBA_U:

If the NAF is able to work with GBA_U keys, i.e., Ks_int_NAF and Ks_ext_NAF, then it would give a GBA_U awareness indication to the BSF. If the GBA_U smart card internal key Ks_int_NAF is intended to be used then the NAF shall send a GSID (GAA Service Identifier) to request the USS identifiable via the GSID. The BSF returns then in the ME key material field the Ks_ext_NAF, since the card is GBA_U-aware. The BSF may include the Ks_int_NAF and the BSF needs to return the USS information if 'Ks_int_NAF' or the 'Ks_int_NAF AND the Ks_ext_NAF' are supposed to be used. The NAF will also get indication that no SIM has been used.

If the NAF is not GBA_U-aware and sends no indicator to the BSF or sets the indicator accordingly, then the BSF returns in the ME key material field only

the Ks_ext_NAF and the UICC key material field would be empty since the NAF
did not indicate that it supports usage of the Ks_int_NAF key. The BSF may
return USS that indicates that Ks_NAF or Ks_ext_NAF is to be used. The NAF
will also get indication that no SIM has been used. If the NAF is not requesting
a USS, then in this case, the NAF is not aware that the key it received is the
Ks_ext_NAF and not the Ks_NAF key.

It should be noted that the support of USS is optional for the operator, but if
an operator wants to use GBA_U, then USS support is required. For GBA_U
the BSF needs to modify the authentication vectors received from HSS and needs
to deduct the information when to make this modification from the USS. If the
card is GBA_U-aware, then the terminal will utilize GBA_U. If a GBA_U-
unaware UICC contains a USIM and a SIM application, then the terminal will
use GBA_ME and interacts with the USIM application for GBA.

3.3.5 3GPP2 GBA

GAA core functionality has also been specified in 3rd Generation Partnership
Project 2 (3GPP2), the North American counterpart of 3GPP. 3GPP2 has agreed to
use GAA specified in 3GPP when the bootstrapping is based on SIM, USIM or ISIM.
However, they have also specified in [S.S0109-0] a way to utilize Code Division
Multiple Access–based (CDMA-based) authentication mechanism in GAA as well.
CDMA is the North American alternative to GSM. These CDMA-based alternatives
for GAA are based on the CDMA2000 1× and CDMA2000 1× EV-DO (Evolution –
Data Only or Data Optimized). CDMA2000 1× is the first phase of CDMA2000.
CDMA2000 1× EV-DO is optimized CDMA carrier for packet data services and the
wireless data access based on CDMA. 3GPP2 bootstrapping mechanisms are based
on password protected Diffie-Hellman key exchange [C.S0016-C] is where the
authentication is done based on CDMA authentication mechanisms. The details and
parameters for Diffie-Hellman key agreement are defined in [S.S0109-0]. It uses the
p value taken from second Oakley group defined in [RFC 2409] but with a generator
g of 13.

CDMA devices are typically equipped with User Identity Module (UIM) or
Removable User Identity Module (R-UIM). The main difference between the two is
that the UIM is part of the terminal and the R-UIM can be physically removed from
the terminal. The R-UIM can be either a stand-alone module as defined [C.S0023-C],
or a multi-application platform, i.e., UICC, that may hold several applications that
can be operated concurrently. In this section, we call all type of UIMs that are
unaware of the GAA, i.e., do not have any GAA-specific functionality *legacy UIMs*.
The UIMs that are aware of GAA are called *GAA-aware UIMs*.

In this section, we describe these legacy CDMA GAA bootstrapping mechanisms
in high level. The reader is referred to 3GPP2 specifications and [S.S0109-0] in

particular to find out the details. For example, how to calculate the BS_RESULT, MS_RESULT, and temporary AKA key can be found in Section 4.5.2.2.3 of [S. S0109-0]. The same section also describes how CK, IK and RES are obtained from the temporary AKA key and the AKA Challenge.

Note about standardization status in 3GPP2: At the time of writing this book, the GAA specification work is currently ongoing in 3GPP2. For example, the Zn interface is currently being standardized and most likely it is based on the work done in 3GPP. Additionally, the specification [S.S0109-0] contains a few flaws that should be noted and corrected:

- 'GET mistake': All of the HTTP requests issued over the Ub interface are GET requests. However, they should be POST requests as the terminal is sending data to the BSF in the HTTP payload.
- 'qop mistake': The specification mistakenly claims that the 401 authentication challenge messages are integrity protected, when qop is set to 'auth-int'. However, as specified in [RFC2617], the challenge message (401) is not integrity protected even though the qop is set to 'auth-int'. This is because in the normal HTTP digest case, the qop in this phase is used to indicate to the client what qop options are supported and allowed by the server. The HTTP payload or any other part of the HTTP response is not protected by the HTTP digest as is hinted in the specification.

The message sequence flows are described in this section as they are specified in [S.S0109-0], thus, the above mistakes are either written in (GET mistake) or omitted (qop mistake).

Bootstrapping with CDMA2000 1×

CDMA2000 1× uses Cellular Authentication and Voice Encryption (CAVE) procedures to authenticate terminals in CDMA2000 1× networks. More details on CAVE authentication procedures can be found in [C.S0023-C], Section 4.2.2. The GAA uses this infrastructure to establish a shared secret between the terminal and the BSF the following way.

The BSF uses the authentication request (AUTHREQ) transactions over MAP protocol [X.S0004-540-E] to request authentication response (AUTHU) and random variable (RAND) unique challenge pair. They are transformed into a RAND and authentication response (AUTHR) pair, and these are used by the BSF to request the associated session keys, i.e., the signalling message encryption key (SMEKEY) and the CDMA private long code mask (CDMAPLCM) from the Foreign Location Registry / Authentication Center (FLR/AC). The RAND parameter (used in CAVE) is transported to the Mobile Node (MN) using HTTP Digest authentication mechanism. The MN will then generate the AUTHR, SMEKEY and CDMAPLCM by sending the RAND to the legacy User Identity Module (UIM) that contains the CAVE functionality. After this, both the BSF and the MN have AUTHR and session keys (SMEKEY and CDMAPLCM) available.

The Diffie-Hellman key agreement is used to ensure the cryptographic separation between the CAVE-generated key material (SMEKEY, CDMAPLCM and AUTHR) and the GAA master session key (Ks). The generated key material is used as the password for authenticating the Diffie-Hellman key agreement between the MN and the BSF. The password (PW) is formed by concatenating the key material: SMEKEY|CDMAPLCM|AUTHR (| denotes the concatenation). These are also noted in the sequence flow below as mobile station password (MS_PW) and base station password (BS_PW), which are equal. During the procedure, the BSF also generate an AKA challenge that is sent to the MN.

The MN and the BSF will utilize the password generated from the PW and the Diffie-Hellman key agreement procedure to generate a temporary AKA key. This temporary key is used as the key (K) in the following AKA functions, the AKA challenge is used as the RAND in standard 3GPP2 AKA functions f3, f4 and f2 functions to generate the CK, IK and RES [S.S0055-A]. These are then used as defined in GBA_ME, i.e., RES is used as the password in HTTP Digest calculations and the GAA master session key (Ks) is generated by concatenating CK and IK. (See Figure 3.10.)

Bootstrapping with CDMA2000 1× EV-DO and Legacy UIM

The CDMA2000 1× EV-DO (Evolution – Data Only / Data Optimized) uses the Mobile IP (MIP) authentication to authenticate terminals in CDMA2000 1× EV-DO networks. MIP authentication procedures can be found in Section 4.7.3 of [C.S0023-C]. The GAA uses this infrastructure to establish a shared secret between the terminal and the BSF the following way.

The BSF and the terminal use HTTP Digest to transport the MIP authentication message exchanges between them. Once the GBA module in the terminal and the BSF possess the MN- Authentication, Authorization and Accounting (MN-AAA) Authenticator, they are ready to perform the password-protected Diffie-Hellman key exchange similarly as in the CDMA2000 1× case in previous section. The communication between the BSF and the Home-AAA (H-AAA) is based on the Remote Authentication Dial-In User Service (RADIUS) protocol. (See Figure 3.11.)

Bootstrapping with CDMA2000 1× EV-DO and GAA-aware UIM

3GPP2 have also defined a case where a CDMA1× EV-DO terminal is equipped with a GAA-aware UIM. In this case, part of the GAA functionality, i.e., key derivation is moved to UIM.

The GAA-aware UIM prevents the usage of normal MIP authentication procedure, which is used with legacy UIMs. This is achieved so that the BSF and the MN derive MS_CHALLENGE* and BS_CHALLENGE* by concatenating the exchanged MS_CHALLEGE and BS_CHALLENGE (i.e., MS_CHALLENGE* = BS_CHAL-LENGE* = MS_CHALLENGE || BS_CHALLENGE). The BSF will send the

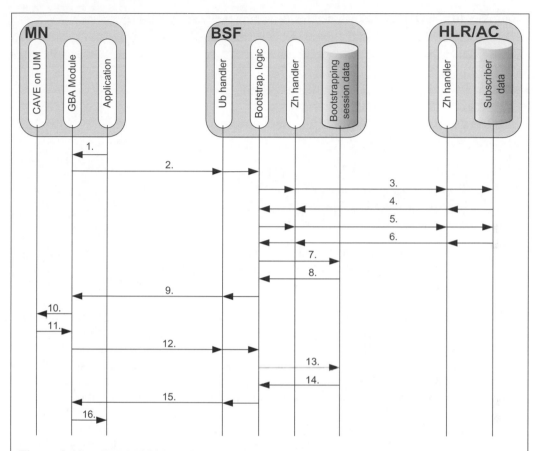

Figure 3.10. CDMA2000 1×-based bootstrapping with legacy UIM

1. An application in the Mobile Node (MN) has come to a state that it needs application-specific GAA keys from the GBA module in the MN. Thus, it contacts the GBA module in the MN and gives it the NAF_ID.
2. The MN sends an HTTP GET request to the BSF, which contains the user's identity in the form of 'IMSI@realm.com' as the username in the Authorization header. The GBA module has obtained this identity earlier from the legacy UIM. The Electronic Serial Number (ESN) value of the MN is also sent to the BSF in the HTTP payload (in CDMA, the ESN is one cornerstone of the general security architecture).

```
GET / HTTP/1.1
Host: bsf.home1.net:80
User-Agent: Bootstrapping Client Agent; 3GPP2 GBA
version 1.0
Date: Thu, 21 Nov 2007 14:09:15 GMT
Accept: */*
Content-Length: 95
```

Figure 3.10.   CDMA2000 1×-based bootstrapping with legacy UIM (*continued*)

```
Content-Type: application/vnd.3gpp2.bsf+xml
Authorization: Digest
        username="<IMSI>@<realm.com>",
        realm="bsf.home1.net",
        nonce="",
        uri="/",
        response=""

<?xml version="1.0" encoding="UTF-8"?>
<BootstrappingInfo xmlns="uri:3gpp2-gba">
  <esn>base64(ESN)</esn>
</bootstrappingInfoType>
```

3. The BSF extracts the IMSI from the username parameter and sends an AUTHREQ message for getting the RANDU and the AUTHU pair to the HLR/AC. The AUTHREQ includes the parameters IMSI, ESN and the SYSACCTYPE set to GBA access.

4. The HLR/AC responds with an authreq, which includes the RANDU and AUTHU pair. The BSF takes the 24-bit RANDU value and concatenates it with the 8 least significant bits of IMSI_S2 derived from the IMSI to create the RAND. The BSF sets the AUTHR equal to the AUTHU. It also generates a secret random number x and calculates (g^x mod p).

   An IMSI_S2 consists out of the most significant 3-digit (10 bits) part of a 10-digit IMSI_S. An IMSI_S is a 10-digit (34 bits) number derived from the IMSI. If an IMSI with equal or more then 10 digits is used, then the IMSI_S is equal to the last 10 digits of the IMSI. If the IMSI has strictly less than 10 digits, then the least significant digits of the IMSI_S are the ones of the IMSI and then to the most significant side zeros are added to obtain 10 digits.

5. The BSF sends an AUTHREQ message for getting the session keys (SMEKEY and CDMAPLCM) from the HLR/AC. The AUTHREQ includes the parameters AUTHR, RAND and the SYSACCTYPE set to GBA access. The SYSCAP parameter is set to indicate that the authentication parameters were requested on the system access (bit A = 1) and that Signalling Message Encryption and Voice Privacy are supported by the system (bit B = 1 and bit C = 1). All other SYSCAP parameters are set to zero.

   The Home Location Register / Authentication Centre (HLR/AuC) validates the AUTHR and generates the SMEKEY and CDMAPLCM.

6. The HLR/AuC transfers the session keys (SMEKEY and CDMAPLCM) to the BSF. Upon receiving the session keys, the BSF calculates the

$$BS\_PW = MS\_PW - SMEKEY|CDMAPLCM|AUTHR.$$

Figure 3.10. CDMA2000 1×-based bootstrapping with legacy UIM (*continued*)

   The BSF also generates a 128 bit random AKA challenge.

7–8. The BSF stores the session data to its local session database.

  9. The BSF sends an HTTP 401 response to the MN. The AKA challenge and RAND are based64 encoded and carried in the nonce field of the WWW-Authenticate header. The payload carries the BS_RESULT parameter.

```
HTTP/1.1 401 Unauthorized
Server: Bootstrapping Server; Release-6
Date: Thu, 21 Nov 2007 14:09:16 GMT
Content-Length: 99
Content-Type: application/vnd.3gpp2.bsf+xml
WWW-Authenticate: Digest
        realm="bsf.home1.net",
        nonce="base64(AKA challenge + RAND)",
        algorithm=AKAv1-MD5,
        qop="auth-int",
        opaque="5ccc069c403ebaf9f0171e9517f30e41"

<?xml version="1.0" encoding="UTF-8"?>
<BootstrappingInfo xmlns="uri:3gpp2-gba">
   <bs_result>base64(BS_RESULT)</bs_result>
</bootstrappingInfoType>
```

10. The GBA module in the MN verifies that the received BS_RESULT is not zero. The RAND challenge value is sent to the legacy UIM as a simulated Global Challenge.

11. The legacy UIM responds to the global challenge with an AUTHR, SMEKEY and CDMAPLCM.

   The GBA module calculates MS_PW / BS_PW the same way as the BSF did in step 6. It then recovers the BS_RESULT received from the BSF in step 9. It then generates a secret random number y for the Diffie-Hellman and calculates (g^y mod p) and (g^{xy} mod p). It then generates the 128-bit random number, CRAND, and calculates the temporary AKA key and the MS_RESULT value. Using the temporary AKA key as K and the AKA challenge (received from the BSF in step 9) as RAND with standard AKA functions, the GBA module can generate CK, IK and a 128-bit RES. The GBA module sets the GAA master session key to be Ks = CK|IK.

12. The MN sends another HTTP GET request to the BSF with an appropriate Authorization header. The Digest response is computed as specified in [RFC2617] using the RES as password. The HTTP request contains CRAND in the cnonce field of the Authorization header and MS_RESULT in the HTTP payload. They are both base64-encoded.

Figure 3.10. CDMA2000 1×-based bootstrapping with legacy UIM (*continued*)

```
GET / HTTP/1.1
Host: bsf.home1.net:80
User-Agent: Bootstrapping Client Agent; Release-6
Date: Thu, 21 Nov 2007 14:09:16 GMT
Content-Length: 99
Content-Type: application/vnd.3gpp2.bsf+xml
Accept: */*
Authorization: Digest
        username="<IMSI>@<realm.com>",
        realm="bsf.home1.net",
        nonce="base64(AKA challenge + RAND)",
        uri="/",
        qop=auth-int,
        nc=00000001,
        cnonce="base64(CRAND)",
        response="6629fae49393a05397450978507c4ef1",
        opaque="5ccc069c403ebaf9f0171e9517f30e41",
        algorithm=AKAv1-MD5

<?xml version="1.0" encoding="UTF-8"?>
<BootstrappingInfo xmlns="uri:3gpp2-gba">
   <ms_result>base64(MS_RESULT)</ms_result>
</bootstrappingInfoType>
```

The BSF verifies that the received MS_RESULT is not zero. The BSF extracts the CRAND from the cnonce field. It will then use the MS_PW / BS_PW to recover (g^y mod p). From this it calculates (g^{xy} mod p) and the temporary AKA key. The BSF will use the temporary AKA key and the AKA challenge the same way as the MN did in step 11 to obtain CK, IK and RES. The BSF then authenticates the MN by verifying the digest response using the RES as the password. If the verification is successful, the BSF sets GAA master session key Ks to be CK|IK.

13–14. The BSF generates the B-TID by taking the base64-encoded AKA challenge value and the BSF server name:

> B-TID = base64encode(AKA challenge)@<BSFservername>.

It then stores the B-TID, Ks and any other implementation-specific parameters to its local session database.

Figure 3.10. CDMA2000 1×-based bootstrapping with legacy UIM (*continued*)

15. The BSF sends a 200 OK response to the MN. The payload contains B-TID and key lifetime. The Authentication-Info header is calculated using the RES as the password and with qop set to 'auth-int', i.e., the HTTP payload is included in the digest calculations and is hence integrity protected as specified in [RFC2617].

```
HTTP/1.1 200 OK
Server: Bootstrapping Server; Release-6
Authentication-Info: qop=auth-int,
        rspauth="6629fae49394a05397450978507c4ef1",
        cnonce="base64(CRAND)",
        nc=00000001,
        opaque="5ccc069c403ebaf9f0171e9517f30e41",
        nonce="base64(AKA challenge + RAND)",
        qop=auth-int
Date: Wed, 21 Nov 2007 14:09:16 GMT
Expires: Wed, 21 Oct 2007 20:09:16 GMT
Content-Type: application/vnd.3gpp2.bsf+xml
Content Length: 185

<?xml version="1.0" encoding="UTF-8"?>
<BootstrappingInfo xmlns="uri:3gpp2-gba">
  <btid>base64(AKA challenge)@bsf.operator.com</btid>
  <lifetime>2007-11-21T20:09:16Z</lifetime>
</BootstrappingInfo>
```

The GBA module verifies digest calculations of Authentication-Info header. If the verification is successful, the BSF is authenticated, and the GBA module stores the Ks, the B-TID, the key lifetime and any other implementation-specific data to its local session cache.

This concludes the bootstrapping procedure based on CDMA2000 1× CAVE procedure.

16. The GBA module derives the application-specific GAA key using the Ks and the NAF_ID. Then returns the application-specific GAA key, B-TID, and key lifetime to the application, and the application can start to use the key with a NAF.

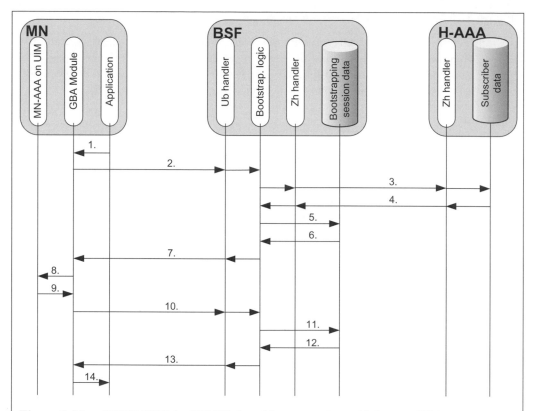

Figure 3.11. CDMA2000 1× EV-DO-based bootstrapping with legacy UIM

1. An application in the Mobile Node (MN) has come to a state that it needs application-specific GAA keys from the GBA module in the MN. Thus, it contacts the GBA module in the MN and gives it the NAF_ID.
2. The MN generates a 16-byte random number, MS_CHALLENGE. The MN sends an HTTP GET request to the BSF, which contains the user's identity in the form of 'IMSI@realm.com' as the username in the Authorization header. The GBA module has obtained this identity earlier from the legacy UIM. The MS_CHALLENGE value is also sent to the BSF in the HTTP payload.

```
GET / HTTP/1.1
Host: bsf.home1.net:80
User-Agent: Bootstrapping Client Agent; 3GPP2 GBA
version 1.0
Date: Thu, 22 Nov 2007 15:09:15 GMT
Accept: */*
Content-Length: 95
Content-Type: application/vnd.3gpp2.bsf+xml
Authorization: Digest
        username="<IMSI>@<realm.com>",
```

Figure 3.11. CDMA2000 1× EV-DO-based bootstrapping with legacy UIM (*continued*)

```
        realm="bsf.home1.net",
        nonce="",
        uri="/",
        response=""


<?xml version="1.0" encoding="UTF-8"?>
<BootstrappingInfo xmlns="uri:3gpp2-gba">
   <ms_chall>base64(MS_CHALLENGE)</ms_chall>
</bootstrappingInfoType>
```

3. The BSF extracts the IMSI from the username parameter. It also generates two random numbers: 128-bit AKA Challenge and a 16-byte BS_CHALLENGE. The BSF also generates a random number x and calculates (g^x mod p). The BSF will then send an Access Request message acting as a RADIUS client to the H-AAA to request the MN-AAA Authenticator associated with the MN. The request contains both MS_CHALLENGE and BS_CHALLENGE parameters.

   The H-AAA calculates the MN-AAA Authenticator using its subscriber database and the MS_CHALLENGE and the BS_CHALLENGE parameters that it received from the BSF.

4. The H-AAA responds with the Access Accept message containing the MN-AAA Authenticator, and the BSF sets the BS_PW and the MS_PW to equal the MN-AAA Authenticator.

5–6. The BSF stores the session data to its local session database.

7. The BSF sends an HTTP 401 response to the MN. The AKA challenge and BS_CHALLENGE are based64-encoded and carried in the nonce field of the WWW-Authenticate header. The payload carries the BS_RESULT parameter.

```
HTTP/1.1 401 Unauthorized
Server: Bootstrapping Server; Release-6
Date: Thu, 22 Nov 2007 15:09:16 GMT
Content-Length: 99
Content-Type: application/vnd.3gpp2.bsf+xml
WWW-Authenticate: Digest
        realm="bsf.home1.net",
        nonce="base64(AKA challenge + BS_CHALLENGE)",
        algorithm=AKAv1-MD5,
        qop="auth-int",
        opaque="5ccc069c403ebaf9f0171e9517f30e41"


<?xml version="1.0" encoding="UTF-8"?>
<BootstrappingInfo xmlns="uri:3gpp2-gba">
   <bs_result>base64(BS_RESULT)</bs_result>
</bootstrappingInfoType>
```

Figure 3.11.   CDMA2000 1× EV-DO-based bootstrapping with legacy UIM (*continued*)

8. The GBA module in the MN verifies that the received BS_RESULT is not zero. The GBA module extracts the BS_CHALLENGE from the nonce parameter in WWW-Authenticate header and sends it together with the previously generated MS_CHALLENGE parameter to the legacy UIM and requests the MN-AAA algorithm in the legacy UIM to compute the MN-AAA Authenticator.

9. The MN-AAA algorithm in the legacy UIM computes the 16-byte MN-AAA Authenticator using the MN-AAA Key, and received MS_CHALLENGE (as Mobile IP Registration request message, MIP-RRQ) and BS_CHALLENGE (as Mobile_IP Challenge), see [C.S0023-C] Section 4.7.3 for details. The MN-AAA Authenticator is returned to the GBA module.

   The GBA module sets MS_PW / BS_PW to be the MN-AAA Authenticator (as the BSF did in step 4). It then recovers the BS_RESULT received from the BSF in step 7. It then generates a secret random number **y** for the Diffie-Hellman and calculates ($g^y \bmod p$) and ($g^{xy} \bmod p$). It then generates the 128-bit random number, CRAND, and calculates the temporary AKA key and the MS_RESULT value. Using the temporary AKA key as K and AKA challenge (received from the BSF in step 7) as RAND with standard AKA functions, the GBA module can generate CK, IK and a 128-bit RES. The GBA module sets the GAA master session key Ks to be CK|IK.

10. The MN sends another HTTP GET request to the BSF with an appropriate Authorization header. The Digest response is computed as specified in [RFC2617] using the RES as password. The HTTP request contains CRAND in the cnonce field of the Authorization header and MS_RESULT in the HTTP payload. They are both base64-encoded.

```
GET / HTTP/1.1
Host: bsf.home1.net:80
User-Agent: Bootstrapping Client Agent; Release-6
Date: Thu, 22 Nov 2007 15:09:16 GMT
Content-Length: 99
Content-Type: application/vnd.3gpp2.bsf+xml
Accept: */*
Authorization: Digest
        username="<IMSI>@<realm.com>",
        realm="bsf.home1.net",
        nonce="base64(AKA challenge + BS_CHALLENGE)",
        uri="/",
        qop=auth-int,
        nc=00000001,
        cnonce="base64(CRAND)",
        response="6629fae49393a05397450978507c4ef1",
        opaque="5ccc069c403ebaf9f0171e9517f30e41",
        algorithm=AKAv1-MD5
```

Figure 3.11. CDMA2000 1× EV-DO-based bootstrapping with legacy UIM (*continued*)

```
<?xml version="1.0" encoding="UTF-8"?>
<BootstrappingInfo xmlns="uri:3gpp2-gba">
   <ms_result>base64(MS_RESULT)</ms_result>
</bootstrappingInfoType>
```

The BSF verifies that the received MS_RESULT is not zero. The BSF extracts the CRAND from the cnonce field. It will then use the MS_PW / BS_PW to recover (g^y mod p). From this, it calculates (g^{xy} mod p) and the temporary AKA key. The BSF will then use temporary AKA key and AKA challenge the same way as the MN did in step 9 to obtain CK, IK and RES. The BSF then authenticates the MN by verifying the digest response using the RES as the password. If the verification is successful, the BSF sets the Ks to be CK|IK.

11–12. The BSF generates the B-TID by taking the base64-encoded AKA challenge value and the BSF server name:

B-TID = base64encode(AKA challenge)@<BSFservername>.

It then stores the B-TID, the Ks, and any other implementation-specific parameters to its local session database.

13. The BSF sends a 200 OK response to the MN. The payload contains B-TID and key lifetime. The Authentication-Info header is calculated using the RES as the password and with qop set to 'auth-int', i.e., the HTTP payload is included in the digest calculations and is hence integrity-protected as specified in [RFC2617].

```
HTTP/1.1 200 OK
Server: Bootstrapping Server; Release-6
Authentication-Info: qop=auth-int,
        rspauth="6629fae49394a05397450978507c4ef1",
        cnonce="base64(CRAND)",
        nc=00000001,
        opaque="5ccc069c403ebaf9f0171e9517f30e41",
        nonce="base64(AKA challenge + BS_CHALLENGE)",
        qop=auth-int
Date: Thu, 22 Nov 2007 15:09:16 GMT
Expires: Thu, 22 Oct 2007 21:09:16 GMT
Content-Type: application/vnd.3gpp.bsf+xml
Content-Type: application/vnd.3gpp2.bsf+xml
Content-Length: 185

<?xml version="1.0" encoding="UTF-8"?>
<BootstrappingInfo xmlns="uri:3gpp2-gba">
   <btid>base64(AKA challenge)@bsf.operator.com</btid>
   <lifetime>2007-11-22T21:09:16Z</lifetime>
</BootstrappingInfo>
```

Figure 3.11. CDMA2000 1× EV-DO-based bootstrapping with legacy UIM (*continued*)

> The GBA module verifies digest calculations of Authentication-Info header. If the verification is successful, the BSF is authenticated, and the GBA module stores the Ks, the B-TID, the key lifetime and any other implementation-specific data to its local session cache.
>   This concludes the bootstrapping procedure based on CDMA2000 1× EV-DO Mobile IP authentication procedure.
> 14. The GBA module derives the application-specific GAA key using the Ks and the NAF_ID. Then returns the application-specific GBA key, B-TID, and key lifetime to the application, and the application can start to use the key with an NAF.

MS_CHALLENGE* and BS_CHALLENGE* to the H-AAA, which retrieves the MN-AAA key of the user, and computes the MN-AAA Authenticator.

The GAA-aware UIM will similarly compute the MN-AAA Authenticator, which will not leave the UIM. Instead, the UIM will return the hash of the MN-AAA authenticator to the ME. Furthermore, GAA-aware UIM will not respond to a normal MN-AAA challenge where the hash of MIP-RRQ is the same as hash of BS_CHALLENGE. This will effectively protect the MN-AAA authenticator as it is never given to the MN in this case as MS_CHALLENGE* and BS_CHALLENGE* are the same.

Other procedures are identical to the bootstrapping case with a legacy UIM. (See Figure 3.12.)

3.4 Building Blocks of GAA

3.4.1 Introduction

The GAA provides a general-purpose service that creates application-specific shared keys between application servers and the mobile terminal. This is a cornerstone on top of which security of various applications and services can be built. How this is done (or should be done) depends on the characteristics of the application but it is also clear that there are many similar needs for security mechanisms between seemingly different services. Therefore, it makes sense to have some general-purpose security services that can be built on top of GAA. These building blocks can then be used in combination with GAA to secure variety of services.

There is always the option to define application-specific mechanism to use GAA-generated keys for the security of the application. This is basically the approach that is taken, e.g., by MBMS. However, this is far from trivial task even in the case it has been done many times before. The approach taken by the building blocks of

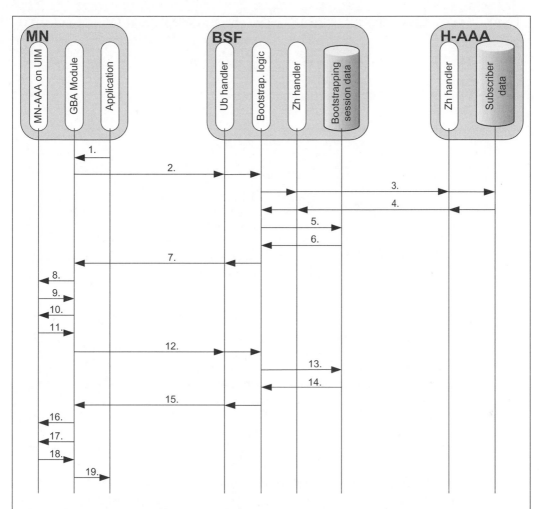

Figure 3.12. CDMA2000 1× EV-DO-based bootstrapping with GAA-aware UIM

1. An application in the MN has come to a state that it needs application-specific GAA keys from the GBA module in the MN. Thus, it contacts the GBA module in the MN and gives it the NAF_ID.
2. The MN generates a 16-byte random number, MS_CHALLENGE. The MN sends an HTTP GET request to the BSF, which contains the user's identity in the form of 'IMSI@realm.com' as the username in the Authorization header. The GBA module has obtained this identity earlier from the legacy UIM. The MS_CHALLENGE value is also sent to the BSF in the HTTP payload.

```
GET / HTTP/1.1
Host: bsf.home1.net:80
User-Agent: Bootstrapping Client Agent; 3GPP2 GBA
version 1.0
```

Figure 3.12. CDMA2000 1× EV-DO-based bootstrapping with GAA-aware UIM (*continued*)

```
Date: Thu, 22 Nov 2007 17:09:15 GMT
Accept: */*
Content-Length: 95
Content-Type: application/vnd.3gpp2.bsf+xml
Authorization: Digest
        username="<IMSI>@<realm.com>",
        realm="bsf.home1.net",
        nonce="",
        uri="/",
        response=""

<?xml version="1.0" encoding="UTF-8"?>
<BootstrappingInfo xmlns="uri:3gpp2-gba">
   <ms_chall>base64(MS_CHALLENGE)</ms_chall>
</bootstrappingInfoType>
```

3. The BSF extracts the IMSI from the username parameter. It also generates two random numbers: 128-bit AKA Challenge and a 16-byte BS_CHALLENGE. The BSF also generates a random number x and calculates (g^x mod p). The BSF discovers that the UIM in the MN is GAA-aware by examining the IMSI and checking it against a local database, where all GAA-aware UIMs identified by IMSIs are stored. Therefore, it will generate MS_CHALLENGE* and BS_CHALLENGE* by concatenating MS_CHALLENGE and BS_CHALLENGE. The BSF will then send an Access Request message acting as a RADIUS client to the H-AAA to request the MN-AAA Authenticator associated with the MN. The request contains both MS_CHALLENGE* and BS_CHALLENGE* parameters.

    The H-AAA calculates the MN-AAA Authenticator using its subscriber database and the MS_CHALLENGE* and the BS_CHALLENGE* parameters that it received from the BSF.

4. The H-AAA responds with the Access Accept message containing the MN-AAA Authenticator, and the BSF sets the BS_PW and the MS_PW to equal the MN-AAA Authenticator.

5–6. The BSF stores the session data to its local session database.

7. The BSF sends an HTTP 401 response to the MN. The AKA challenge and BS_CHALLENGE are based64-encoded and carried in the nonce field of the WWW-Authenticate header. The payload carries the BS_RESULT parameter.

```
HTTP/1.1 401 Unauthorized
Server: Bootstrapping Server; Release-6
Date: Thu, 22 Nov 2007 17:09:16 GMT
Content-Length: 99
Content-Type: application/vnd.3gpp2.bsf+xml
WWW-Authenticate: Digest
```

Figure 3.12. CDMA2000 1× EV-DO-based bootstrapping with GAA-aware UIM (*continued*)

```
                   realm="bsf.home1.net",
                   nonce="base64(AKA challenge + BS_CHALLENGE)",
                   algorithm=AKAv1-MD5,
                   qop="auth-int",
                   opaque="5ccc069c403ebaf9f0171e9517f30e41"


   <?xml version="1.0" encoding="UTF-8"?>
   <BootstrappingInfo xmlns="uri:3gpp2-gba">
     <bs_result>base64(BS_RESULT)</bs_result>
   </bootstrappingInfoType>
```

8. The GBA module in the MN verifies that the received BS_RESULT is not zero. The GBA module extracts the BS_CHALLENGE from the nonce parameter in WWW-Authenticate header. The GBA module is aware that the UIM in the MN is GAA-aware and therefore it will generate the MS_CHALLENGE* and BS_CHALLENGE* as the BSF did in step 3. Then it sends these parameters to the GAA-aware UIM and requests the MN AAA algorithm in the legacy UIM to compute the MN AAA Authenticator.

9. The MN-AAA algorithm in the GAA-aware UIM computes the 16-byte MN-AAA Authenticator using the MN-AAA Key, and received MS_CHALLENGE* (as Mobile IP Registration request message, MIP-RRQ) and BS_CHALLENGE* (as Mobile_IP Challenge), see [C.S0023-C] Section 4.7.3 for details. As the GAA-aware UIM detect that the MS_CHALLENGE* and BS_CHALLENGE* are equal, the GAA-aware UIM will only return the SHA-1 hash of the MN-AAA Authenticator to the GBA module.

    The GBA module uses the hash of the MN-AAA Authenticator to calculate a BS_PW_HASH and a MS_PW_HASH values, which are equal (see details [S.S0109-A]). The BS_PW_HASH is used to obtain (g^x mod p). It then generates a secret random number y for the Diffie-Hellman and calculates (g^y mod p) and (g^{xy} mod p). Finally, the GBA module generates MS_RESULT value using MS_PW_HASH and (g^y mod p).

10. The GBA module sends the AKA challenge (received from the BSF in step 7) and SHA-1 hash of (g^x, g^y, g^{xy}) to the GAA-aware UIM.

11. The GAA-aware UIM generates the 128-bit random number, CRAND, and calculates the temporary AKA key using the CRAND, received hash of (g^x, g^y, g^{xy}), and the stored MN-AAA Authenticator. Using the temporary AKA key as K and AKA challenge (received originally from the BSF in step 7) as RAND with standard AKA functions, the GBA module can generate CK, IK and a 128-bit RES. The GBA module sets the GBA master secret Ks to be CK|IK. The GAA-aware UIM returns only the CRAND and RES to the MN.

12. The MN sends another HTTP GET request to the BSF with an appropriate Authorization header. The Digest response is computed as specified in [RFC2617] using the RES as password. The HTTP request contains CRAND in the cnonce field

Figure 3.12. CDMA2000 1× EV-DO-based bootstrapping with GAA-aware UIM (*continued*)

of the Authorization header, and MS_RESULT (calculated in step 9) in the HTTP payload. They are both base base64 encoded.

```
GET / HTTP/1.1
Host: bsf.home1.net:80
User-Agent: Bootstrapping Client Agent; Release-6
Date: Thu, 22 Nov 2007 17:09:16 GMT
Content-Length: 99
Content-Type: application/vnd.3gpp2.bsf+xml
Accept: */*
Authorization: Digest
        username="<IMSI>@<realm.com>",
        realm="bsf.home1.net",
        nonce="base64(AKA challenge + BS_CHALLENGE)",
        uri="/",
        qop=auth-int,
        nc=00000001,
        cnonce="base64(CRAND)",
        response="6629fae49393a05397450978507c4ef1",
        opaque="5ccc069c403ebaf9f0171e9517f30e41",
        algorithm=AKAv1-MD5

<?xml version="1.0" encoding="UTF-8"?>
<BootstrappingInfo xmlns="uri:3gpp2-gba">
   <ms_result>base64(MS_RESULT)</ms_result>
</bootstrappingInfoType>
```

The BSF verifies that the received MS_RESULT is not zero. The BSF extracts the CRAND from the cnonce field. It will then use the MS_PW / BS_PW to recover (g^y mod p). From this, it calculates (g^{xy} mod p) and the temporary AKA key. The BSF will then use temporary AKA key and AKA challenge the same way as the MN did in step 11 to obtain CK, IK and RES. The BSF then authenticates the MN by verifying the digest response using the RES as the password. If the verification is successful, the BSF sets the Ks to be CK|IK.

13–14. The BSF generates the B-TID by taking the base64-encoded AKA challenge value and the BSF server name:

> B-TID = base64encode(AKA challenge)@<BSFservername>.

It then stores the B-TID, the Ks, and any other implementation-specific parameters to its local session database.

15. The BSF sends a 200 OK response to the MN. The payload contains B-TID and key lifetime. The Authentication-Info header is calculated using the RES as the password and with qop set to 'auth-int', i.e., the HTTP payload is included

Figure 3.12. CDMA2000 1× EV-DO-based bootstrapping with GAA-aware UIM (*continued*)

in the digest calculations and is hence integrity-protected as specified in [RFC2617].

```
HTTP/1.1 200 OK
Server: Bootstrapping Server; Release-6
Authentication-Info: qop=auth-int,
        rspauth="6629fae49394a05397450978507c4ef1",
        cnonce="base64(CRAND)",
        nc=00000001,
        opaque="5ccc069c403ebaf9f0171e9517f30e41",
        nonce="base64(AKA challenge + RAND)",
        qop=auth-int
Date: Wed, 21Thu, 22 Nov 2007 1417:09:16 GMT
Expires: Wed, 21Thu, 22 Oct 2007 2023:09:16 GMT
Content-Type: application/vnd.3gpp.bsf+xml
Content-Type: application/vnd.3gpp2.bsf+xml
Content-Length: 185

<?xml version="1.0" encoding="UTF-8"?>
<BootstrappingInfo xmlns="uri:3gpp2-gba">
   <btid>base64(AKA challenge)@bsf.operator.com</btid>
   <lifetime>2007-11-21T2022T23:09:16Z</lifetime>
</BootstrappingInfo>
```

16. The GBA module verifies digest calculations of Authentication-Info header. If the verification is successful, the BSF is authenticated, and the GBA module stores the GAA master key Ks, the B-TID, and the key lifetime, and any other implementation-specific data to its local session cache, the GAA-aware UIM.
    This concludes the bootstrapping procedure.
17. The GBA module proceeds with the generation of the application-specific GAA key as was requested by the application in step 1. The GBA module sends the NAF_ID and a Network Access ID (NAI) of the GAA-aware UIM to the GAA-aware UIM to request the application-specific GAA key.
18. The GAA-aware UIM derives two keys using the stored Ks, the AKA challenge (= RAND), the NAI, and NAF_ID: Ks_int_NAF that stays the UIM, and Ks_ext_NAF that was given to the MN. The GAA-aware UIM returns the Ks_ext_NAF key to the MN.
19. The GBA module returns the application-specific GAA key (Ks_ext_NAF), the B-TID and the key lifetime to the application, and the application can start to use the key with an NAF.

GAA is to define how certain key security applications can make use of GAA. That means we define the usage of GAA-based keys for these security applications making the task of providing security for the target applications much easier. Basically, with the whole GAA, i.e., GAA bootstrapping and a suitable set of the other building blocks on top of it, we have a handy toolbox that can be used for design of security mechanisms for other applications.

GAA has been introduced in 3GPP Release 6. It has then been extended with further functionality and additional building blocks. The building blocks and further GAA enablers that have been defined for 3GPP GAA include the following:

- Provisioning of subscriber (or client) certificates in a secure way. This feature is called Support for subscriber certificates (SSC) and is specified in [TS33.221]. Many applications have a built-in support for certificates, totally independent of whether these certificates are provided by GAA or some other method. In the GAA terminology, this kind of special NAF is a *PKI portal.*
- Support for HTTPS sessions [TS33.222]. HTTPS refers to a secure HTTP and it has been specified how GAA-generated secrets can be used as credentials for HTTPS. Here, the NAF can be a *web server* or an *authentication proxy* (AP) that provides the HTTPS-secured tunnel for applications that need to be secured.
- Key distribution service for terminals and remote devices [TS33.259]. GAA as such is essentially a key generation service that provides keys between the UE and the network element called NAF. In many cases, it would be beneficial to have a shared key between, e.g., two terminals. In this building block, the NAF is a *key distribution server.* The key distribution service is a feature of Release 7.
- Key distribution service for smart card and terminal [TS33.110]. Similarly as in the previous building block, we get a key generation service to provision keys to the smart card and the terminal (also includes the case that the terminal might be remote device). In this building block, the NAF is a *key distribution server.* Also this type of key distribution service is a feature of Release 7.

3.4.2 PKI Portal

In the following, we describe how the UE is able to fetch a subscriber certificate from the specific NAF, called PKI portal [TS33.221]. We include the case where the UE has a Wireless Identity Module (WIM), which is where the private key is stored, and the corresponding public key being the one that needs to be certified. The private keystore may be also in other format than WIM, and the same protocol usable with other types of keystores like.hardware tokens, and software key repositories. (See Figure 3.13.)

The usage of the subscriber certificates has basically very few limitations. The most usual format of certificates, X.509, is supported by the method. Defining the use cases is mainly out of the scope of 3GPP but maybe done in fora like OMA for certain cases.

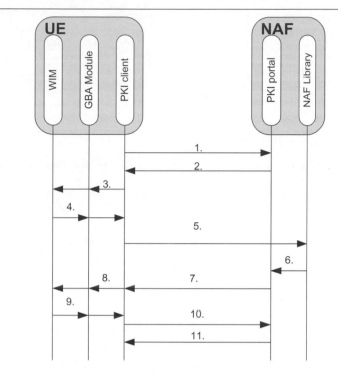

Figure 3.13. PKI Portal (NAF)

1. The PKI application in the UE sends an empty HTTP request to trigger the procedure.
2. The PKI portal responds with HTTP response 401.
3. The GBA module is invoked to obtain B-TID and Ks_NAF. In case proof-of-origin is needed for the WIM, also the WIM is contacted.
4. The parameters are delivered to the PKI application.
5. Based on the parameters obtained from GBA module and WIM, another HTTP request is sent to the PKI portal, this time with the Authorization header.
6. The PKI portal obtains B-TID and Ks_NAF with the help of the NAF library through the Zn interface. With these parameters, the PKI portal verifies the Authorization header.
7. The HTTP response is returned, including the response for the WIM challenge.
8. Further WIM AssuranceInfo is requested from the WIM.
9. WIM provides the parameters that give the needed proof-of-origin. Parameters needed for the PKCS#10 purposes are also delivered.
10. The PKI client sends the PKCS#10 request for a certificate.
11. The PKI portal delivers the certificate in the HTTP 200 OK message.

3.4.3 HTTPS Support

HTTPS is one of the most commonly known usages for security. GAA can be used to support HTTP Digest Authentication and Pre-Shared TLS as outlined in [TS33.222]. The details were already described in Section 3.2.3 (GAA Authentication) and further described in Section 4.1.1.

3.4.4 Key Distribution Service

3.4.4.1 Key Distribution for Terminal to Remote Device Usage

The key distribution service for terminals and remote devices is outlined in [TS33.259]. It is a security feature that builds upon [TS33.220] to provision a shared key between a UICC Hosting Device and a remote device connected via a local interface. This can be used, for example, when a terminal, with a UICC inserted, is connected via local means (cable, Bluetooth, etc.) to a PC or other terminal. The shared secret is then intended to be used to secure the local communication between those two devices.

   The architecture below is based on GAA, but the NAF is a special-purpose NAF. The NAF serves as a key distribution server to the requesting devices. Also the usage of the key is different since it is not used between a network server and a terminal, but between two entities, where one is a terminal with a UICC and the other one might be a terminal, but can also be another entity, e.g., a PC or a personal network device. (See Figures 3.14 and 3.15.)

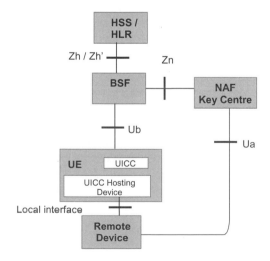

Figure 3.14. Architecture for key distribution service

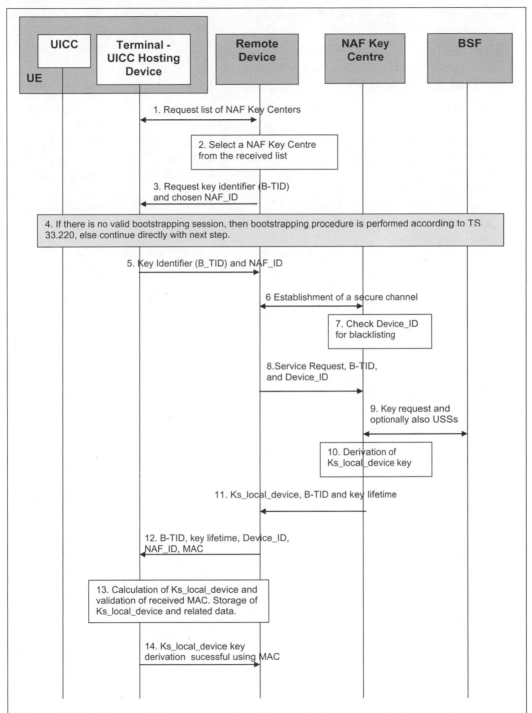

Figure 3.15. Key establishment procedure for terminal to remote device security

Figure 3.15. Key establishment procedure for terminal to remote device security (*continued*)

The Remote Device wants to secure the communication to the Terminal and checks, if it has a valid key for this Terminal (the key is Ks_local_device). In the case that there is no key that it can use, then the following procedure starts:

1. The Remote Device checks for the presence of a list of servers with NAF key distribution servers (NAF Key Centre) functionality. If the Remote Device has already a list and knows which NAF Key Centre to contact, then it can proceed directly with step 3. If no list is available, then it requests the UICC Hosting Device to send a list of supported NAF key distribution servers. The UICC Hosting Device returns then the list of available NAF Key Centres.
2. The Remote Device then chooses an NAF Key Centre from the list (or it may also propose its own NAF Key Centre) and stores the choice for the next time.
3. The Remote Device contacts the UICC Hosting Device, sends the identifier for the chosen NAF Key Centre (NAF_ID) and a requests the key identifier.
4. If the UICC Hosting Device does not have a valid bootstrapping session, then it runs a new bootstrapping procedure with the BSF. The bootstrapping can be based on GBA_U or GBA_ME. When GBA_U has been performed, then the UICC transfers the Ks_ext_NAF key to the UICC Hosting Device. The UICC Hosting Device now has the NAF-specific keys, i.e., either Ks_ext_NAF or Ks_NAF. For this key derivation, the NAF_ID received from the Remote Device is needed.
5. The UICC Hosting Device sends a response to the Remote Device and includes the key identifier B-TID and NAF_ID.
6. Now the Remote Device and the NAF Key Centre establish a secure channel using TLS with certificate-based mutual authentication according to TLS [RFC2246] and its extensions [RFC3546]. A secure channel can also be established using the shared key with the NAF Key Centre and PSK TLS [RFC4279].
7. Through this secure tunnel, the Remote Device sends a 'service request' to the NAF Key Centre to request a device-specific key. The request contains the key identifier (B-TID) received from the UICC Holding Device and an identifier for the Remote Device, called Device_ID, e.g., IMEI. It should be noted that the specification does not address how Device_ID is authenticated.
8. The NAF Key Centre checks if the Device_ID is blacklisted. If it is blacklisted, then the secure tunnel is terminated.
9. The NAF Key Centre requests the NAF-specific key and optionally USSs from the BSF over Zn interface. The BSF returns the requested data, if allowed by BSF policy (see Section 3.5.1).
10. The NAF Key Centre derives the Ks_local_device from the received Ks_(ext)_NAF key, B-TID, NAF_ID and Device_ID. Ks_local_device is then stored.

Figure 3.15. Key establishment procedure for terminal to remote device security (*continued*)

11. The NAF Key Centre sends the Ks_local_device together with the B-TID and the key lifetime through the secure tunnel to the Remote Device. The Remote Device stores the received data for further usage.
12. The Remote Device sends the NAF_ID, Device_ID, B-TID, key lifetime, and a MAC, where MAC is calculated as follows:

$$MAC = \text{HMAC-SHA256 (Ks\_local\_device, NAF\_ID} \parallel \text{Device\_ID\_} \parallel \text{B\_TID)},$$
truncated to 16 octets.

The UICC Hosting Device knows then that the key establishment was successful.
13. The UICC Hosting Device derives the Ks_local_device key and validates the MAC. The Ks_local_device key and the related data (e.g. Device_ID) are then stored.
14. The UICC Hosting Device sends a confirmation of the successful derivation to the Remote device using a MAC and the Ks_local_device key.

The trustworthiness of the devices can be assured e.g. if they comply to the TCG (Trusted Computing Group) MPWG (Mobile Phone Working Group) Mobile Phone Specifications [TCG] or to other TCG technology.

3.4.4.2 Key Distribution for UICC to Terminal Usage

The previous section described how to provision keys to secure the communication between a terminal and another device. Now we will look into the same problem for the case that a security association is needed between a UICC and a terminal. The situation is very similar, but there are small subtle differences due to the fact that the UICC is more closely related to an operator than a terminal. The mechanism is specified in [TS33.110].

The architecture is the same as in the previous section, but the goal is to secure the link between the UICC and the UICC hosting device and not (as in the previous section) to secure the local link. Compared to the previous section, roughly the terminal takes the role of the remote device and the UICC takes the role of the UICC hosting device. The trustworthiness of the terminal can be guaranteed for example by complying with the TCG specifications [TCG]. The message flow is as follows:

1. The terminal contacts the UICC to check if there is a valid master key (Ks) in the UICC. This is done by requesting the key identifier (B-TID) and the key lifetime from the UICC. If there is no valid master key, then the terminal runs a GBA_U bootstrapping procedure.
2. The terminal sends the B-TID and the NAF_ID of the NAF Key Centre to the UICC to determine if there is a valid NAF Key Centre-specific key, i.e., Ks_int_

NAF. The NAF_ID of the NAF Key Centre might be first retrieved from the UICC.

3. The terminal and the NAF Key Centre establish a secure channel using certificate-based mutual authentication [TS33.222], or PSK TLS [RFC4279].

4. Through this secure tunnel, the terminal sends a 'service request' to the NAF Key Centre in the network containing the key identifier (B-TID), the terminal identifier (Terminal_ID), the smart card identifier – Integrated Circuit Card ID (ICCID), the application identifier of the UICC (UICC_appl_ID) and the terminal application identifier (Terminal_appli_ID). The request triggers the establishment of the Ks_local key and a RANDx value (which can be random or a timestamp or something else produced by the terminal). It should be noted that due to the send parameters, the Ks_local key depends on the terminal and the smart card application, i.e., a different application would get a different key. There is the possibility to generate the key on a per platform basis by setting UICC_appli_ID and Terminal_appli_ID equal to 'platform'.

5. The NAF Key Centre makes a blacklist check using the received Terminal_ID and then uses the Zh interface to request the GBA_U NAF-specific keys and may also request one or several USSs.

6. The BSF derives the NAF Key Centre-specific Ks_int/ext_NAF keys and sends them with the further key information and USSs to the NAF Key Centre according to the local policy. It is assumed that since the NAF Key Centre resides in the same network as the BSF, then there should be no reason for the BSF not to send the requested USSs.

7. The NAF Key Centre checks the USS (if received) if the user is authorized to use this service. The NAF Key Centre generates a Counter Limit for use in the UICC. Then the NAF Key Centre derives the Ks_local = Key Derivation Function (Ks_int_NAF, B-TID, Terminal_ID, ICCID, Terminal_appli_ID, UICC_appli_ID, RANDx, CounterLimit).

8. Through the secure channel, the NAF Key Centre sends to the Terminal the B-TID, the Ks_local key, the key lifetime and the Counter Limit. The Terminal stores then the data and related information.

9. The Terminal sends a request to the UICC to derive the Ks_local and includes the needed data for key derivation. Additionally, a MAC is included.

10. The UICC then retrieves the B-TID and the Ks_int_NAF (and potentially a local policy) and derives the Ks_local. The UICC derives also the MAC and compares it with the received one. If the comparison was successful, this is indicated to the Terminal and the Ks_local is used.

It should be noted, that the key establishment between a UICC and a terminal requires many new elements for the interface between UICC and terminal. The support of [TS33.110] implies that the operator has to issue special UICC smart cards, which support this specific functionality.

3.5 Other Architectural Issues

In this section, we will briefly discuss some architectural and design issues we have not addressed so far.

3.5.1 Access Control Mechanisms in GAA

GAA uses GUSS for additional access control. This element contains security-relevant information about the subscriber and the services he or she can access. Each subscriber may have a GUSS element in home operator's databases. The GUSS includes zero or more User Security Settings (USS) elements that contain either additional persistent identities or pseudonyms, or so called authorization flags, or both. There is only identity and authorization-related data in the USS as it is intended for security-related decisions such as:

- what is the service-specific and persistent identity that is in use for the subscriber, or
- whether a subscriber is allowed to use a service, or only certain parts of the service?

Any other data, e.g., user profiles should be either stored in the NAF itself, or the NAF can retrieve the other user data from some other server.

When NAF requests the information from the BSF, it can request zero or more USS elements to be included to the reply. If appropriate USSs exist in subscriber's GUSS, the BSF will include them to the reply provided that the NAF has been authorized to receive them. The existence of particular USS in subscriber's GUSS can also be used as an access control mechanism in the BSF, i.e., the BSF may have been configured so that for particular NAFs, the subscriber must have a certain USS. Otherwise, the subscriber is not allowed to access the service offered by the NAF. This functionality enables the MNO to control what services its subscribers have access to.

GUSS has also a BSFInfo element that contains security-related information intended for the BSF and this element is never given to the NAFs. It contains information such as:

- uiccType, which indicates what kind of smart card the subscriber has, i.e., is it GBA-aware (GBA_U) card or not; and

- lifeTime, which can be used to set a subscriber-specific lifetime for the GAA keys, and override the default key lifetime setting in the BSF.

The BSFInfo element may also contain extensions which the operator or BSF vendor can additionally define for further enhancing the GAA functionality in the BSF and in other GAA nodes.

3.5.1.1 Local Policy Enforcement in the BSF

Local policies in the BSF are operator-specific, but operators can utilize USSs for communicating those policies from the HSS to the BSF, and further to the NAF. Depending on the different capabilities of the BSF and the NAFs, the following scenarios are possible:

(1) The **BSF does not have** a local policy for this NAF and the **NAF is not using** USSs, and therefore, does not request one or several USS from the BSF, i.e., no GSID(s) are inserted.
(2) The **BSF does have** a local policy for this NAF, but the **NAF is not using** USSs.
(3) The **BSF does not have** a local policy for this NAF, but the **NAF is using** USSs and requests USSs from the BSF.
(4) The **BSF does have** a local policy for the NAF and the **NAF is using** USSs.

We will now briefly outline what happens in each scenario with regard to the policy enforcement of the home operator hosting the BSF. The NAF has received in all scenarios the Bootstrapping Transaction Identifier (B-TID) from the terminal as outlined in Sections 3.2 and 3.3.

The first scenario is the simplest one; here the BSF does not have a local policy for this NAF and the NAF is not using USSs. Therefore, the NAF only requests the NAF-specific key(s) and does not request one or several USS from the BSF, i.e., no GSID(s) are inserted in the Zn request. The BSF retrieves then the subscriber information from his local database and checks if there is a local policy for this NAF. Since in this scenario the BSF does not have a local policy for this NAF, the BSF does not require that there be a specific USS present in the subscriber's GUSS. The BSF just generates the NAF-specific key(s) and returns those to the NAF without any USSs. Independent of this, the NAF may still have his own local NAF policy that he enforces.

In the second scenario, the BSF does have a local policy for this NAF, but the NAF is not using USSs. The BSF notices that the NAF did not include any GSIDs in his key request message. The BSF retrieves the subscriber information and finds that there is a policy and there are required GSIDs for this NAF. It checks whether the required USSs are present in the GUSS. If they are present in the GUSS, then the BSF can continue and hand out the key(s) to the NAF, else it has to send an error message. Again, the NAF can have additionally his own local policy.

In the third scenario, the BSF does not have a local policy for this NAF, but the NAF is using USSs and requests USSs from the BSF. This could be the case when, for example, one NAF with a very popular service has security relationships with many operators and some of those have BSF that support local policies and some of those have BSF that do not support. The NAF would most likely not modify its request over Zn interface depending on which operator he is addressing, hence this case has to be considered. The NAF requests the NAF-specific shared key(s) from the BSF and includes one or several GSIDs in this request. The BSF then retrieves the subscriber information and checks if a local policy exist for this NAF. Since there is no local policy for this NAF, the BSF does not require any specific USS to be present in the GUSS requested from the HSS. The BSF generates the keys and sends the key(s) and the requested USS (if available) to the NAF. The NAF then uses the keys and processes the (potentially) received USS.

In the fourth and 'fully enabled' scenario, the BSF does have a local policy for this NAF and the NAF uses USSs. The NAF requests the shared key(s) and the USSs identified by GSIDs. The BSF retrieves the subscriber information from his local database and check if there is a policy for this NAF. Since in this scenario the BSF does have a local policy, the BSF requires that there is one or more specific USSs identified by the GSIDs to be present in the GUSS. If all the required USSs are present in the GUSS, the BSF sends then the received USSs to the NAF, which then processes the USSs and the keys as outlined in the previous section. If even one of the required USSs is missing, the BSF sends an error message to the NAF.

These scenarios are also described in Annex J of [TS33.220]. It should be noted, that for the last two scenarios, the BSF might refrain from sending a USS to a NAF, if the NAF is not authorized to receive it.

3.5.1.2 USS usage for NAFs

The basic need for NAF to use USS is to obtain a persistent identity or pseudonym so that it can provide a user-specific service to the end-user. This can be done without USS by using the IMPI value that may be returned by the BSF to the NAF. However, often the MNO does not want to reveal the IMPI to the NAF unless the NAF is operated by the MNO itself. In this case, the NAF can request a service-specific USS from the BSF by using a GAA Service Identifier (GSID). This may be a standardized value or may be given by the MNO to the NAF. In both cases, the NAF has been configured with GSID value, which it would include to all requests to the BSF. The BSF will return the USS identified by the GSID if the particular subscriber has the USS element and if the NAF authorized to get it. The policy about whether the NAF is authorized to receive it or not is locally configured in the BSF by the MNO. When the NAF gets the USS, it can examine its content and use the provided persistent pseudonym to map the subscriber to a local (or remote) user profile data.

The other need for NAF to use USS is to obtain authorization data from the BSF on whether a subscriber is allowed to use a service (see BSF policy enforcement in the previous section). This can be either explicit or implicit and it can offer authorization information regarding the whole service or only to certain parts of the service.

In the explicit case, the authorization decision is made by the NAF. In this case, a USS has typically been returned to the NAF. If the USS does not contain any authorization flag data, then the NAF can provide the service to the user. If the USS contains authorization flags, then these flags identify the parts of the service that a user either could access or should be denied access.

A service offered by the NAF may be divided into certain service levels, e.g., standard account access and premium account access. In this case, the authorization flags would indicate to the NAF to which account class the subscriber belongs to and the NAF can then offer the service level that subscriber is entitled to. Naturally, if the requested USS does not exist for the subscriber and the policy in the NAF requires it to exist, then the subscriber is not authorized to access the service. Hence, the pure existence of a USS can be used to control access to a service in the NAF.

In the implicit case, the authorization decision is made by the BSF as described in previous section. This is based on whether a particular USS identified by a certain GSID exists in subscriber's GUSS. The BSF may be configured in such a way that it requires that a particular USS must exist in subscriber's GUSS when a certain NAF send a request, regardless of whether it requests this USS or not. If the USS does not exist, then the BSF takes this as an indication that this particular subscriber is not allowed to use the service offered by the NAF and it will just return 'not allowed' message to the NAF as an indication of this. The BSF will not even send the keys to the NAF. If the USS does exist, then the response with GAA keys is sent normally to the NAF. Note that the NAF is not required to request the USS in question.

3.5.2 Identities in GAA

GAA uses and supports multiple identities that can be used in different interfaces of GAA. The long-term persistent identity that GAA uses is the cellular identity; in 2G and 3G cases, this is the International Mobile Subscriber Identity (IMSI), and in the IMS case, it is the IP Multimedia Private Identity (IMPI). These are used to map subscriber data with an identity both in the UE, namely in the UICC and in the HSS. This long-term identity is used during bootstrapping procedures over Ub and Zh interfaces to handle the AKA procedures.

Once a bootstrapping session has been completed, the BSF creates a key B-TID, which the BSF sends to the UE over Ub interface. The key identifier B-TID is used to identify a bootstrapping session between the UE, the BSF and the NAF. If only B-TID is used between the UE and the NAF, then NAF can only be certain that this subscriber has valid subscription with the MNO provided that the authentication over Ua interface is successful.

In the cases where NAF wants to map the subscriber to a persistent user account to provide a user-specific service, it needs to obtain a persistent identifier or identifiers from the BSF. This can be done in two ways:

(a) The BSF may have been configured by the MNO to return the cellular identity (IMPI) to the NAF. This may be the case if the MNO trusts the NAF to handle the private identity.
(b) The NAF may request an application-specific user identities or pseudonyms from the BSF.

In the latter case, the BSF has fetched a subscriber-specific GUSS data element from the HSS during bootstrapping procedure. The GUSS contains application-specific User Security Settings (USS) elements that contain application-specific identities and authorization information (a.k.a. flags) that are identified by a GAA Service Identifiers (GSIDs). The NAF may request one or more USS elements to be sent back together with the application-specific key, and thus use the service identifier to handle the persistent mapping of the user to a local user account, for example.

Figure 3.16 summarizes the different identifiers being used in different interfaces of GAA.

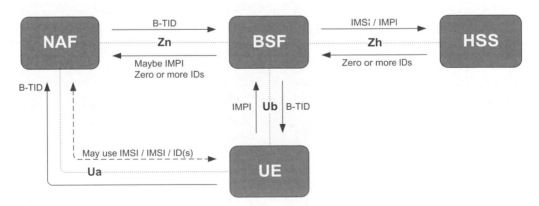

Figure 3.16.   GAA identifier usage

In order to keep the application-specific user identities secret, the BSF has been configured in such a way that only certain NAFs have access to a certain USS elements. If a USS is requested by an NAF that is not authorized to do so, the BSF will not include that USS to the response message.

3.5.3 Identity Privacy and Unlinkability

One of the goals in design of cellular networks has been to protect the identity of mobile users and the data about their movements from unauthorized third parties. The GSM system, for example, supports the notion of Temporary Mobile Subscriber Identifiers (TMSIs), which the UE can use during signalling instead of its real IMSI.

There are two places in GAA where the subscriber identity may be revealed to an observer:

(a) during bootstrapping via the Ub interface, the private user identity, IMPI, is sent in the initial request from UE to BSF; and
(b) during application usage via the Ua interface, the same bootstrapping identifier B-TID is sent in every initial Application Request during the time that a certain bootstrapped master session key is in use.

Without further precautions, an eavesdropper watching the exchanges over the Ub interface will learn the private user identity. An eavesdropper watching the exchanges over the Ua interface will not learn the private user identity, but will be able to link multiple application sessions in which the same UE participated.

The first threat can be addressed by doing bootstrapping via a secure channel. For example, if bootstrapping takes place through a server-authenticated TLS tunnel, there is no risk of leaking the private user identity to onlookers.

The second threat can be addressed similarly: by running the application protocol inside a server-authenticated tunnel, onlookers will not be able to link two different application sessions by the same UE even if they use the same B-TID.

Some NAFs may need to know the subscriber's private user identity. Whether an NAF is allowed to have this information is controlled by policy set by the home operator in GAA User Security Settings (GUSS).

3.5.4 Usability and GAA

The majority of Internet services rely on username and password authentication at the time of writing, and that authentication method is applied in two main ways. In the first, the user device and the server establish an encrypted TLS channel and the network side of that channel is authenticated based on server certificate. The user then fills his username and password into an HTML form, which is sent to the server through the encrypted TLS channel. Mutual authentication succeeds if the username and password in the received HTML form are verified by the server.

Figure 3.17 illustrates the web browser interface of a mobile device during authentication with this method. One detail of this interface is worth mentioning: If the user clicks on 'Remember me' box that is part of the HTML form, the server will set up a HTTP cookie to the browser. In subsequent authentications, that cookie will be automatically sent to the server and used instead of username and password.

Second, username and password are also often used in HTTP-digest authentication. Figure 3.18 illustrates the web browser interface of a mobile device during HTTP-digest authentication.

Another service authentication method, which may not require the user to enter username and password, is to authenticate both TLS channel ends with certificates. The user may need to input a passphrase in order to access the private key associated with client certificate. The network is authenticated with server certificate and the terminal with a client certificate. However, so far, this method has been used rarely compared to username and password. A likely reason is that a large-scale investment in distribution and management of client certificates in user terminals is hard to justify if username and password method is considered good enough.

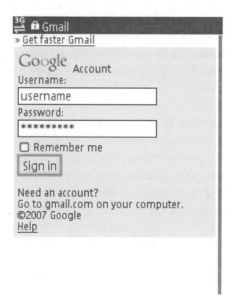

Figure 3.17. HTML form for password authentication in a mobile device. Reproduced by permission of © Nokia

Figure 3.18.   Entering password during HTTP-digest authentication in mobile device. Reproduced by permission of © Nokia

In summary, the main difficulty for the user with the current mainstream authentication mechanisms is the need to create, remember and manage passwords. To mitigate this difficulty, browsers can implement a password manager that if enabled, will save and automatically complete the username and password in future authentications. But this still leaves the password-generation task to the user and people are not good at choosing secure passwords [Yan04].

For services that can rely on cellular authentication, GAA offers a more secure and user-friendly alternative: A service–specific shared secret is automatically generated for each service using GAA and the lifetime of that key is set by the MNO. In the rest of this section, we will discuss an implementation issue that is linked to usability: under what conditions an application can receive that service-specific shared secret transparently to the user.

Cellular services authentication is automatic. As a result, the user does not see dialogues like the ones in Figures 3.17 and 3.18 when placing and answering calls and transferring data over cellular network. We feel that automatic authentication contributes greatly to the ease-of-use feeling that is associated with cellular service. (Another and no less important fact that contributes to the ease-of-use feeling is that cellular service works most of the time.)

Since GAA is based on cellular authentication, it can inherit this usability feature: The GBA module in the device would bootstrap as needed and provide credentials to applications silently, without involving the user. In this case, the decision on asking or not asking the user to confirm authentications is left to the application that gets the bootstrapped credentials. This is how GBA module is implemented, e.g., in Series60 platform.

An issue with silent operation of GBA module is that, unbeknownst to the user, a malicious or faulty application on the mobile device may request and misuse bootstrapped credentials for a service. This issue is mitigated by controlling access of applications to bootstrapped credentials, which can be done in two ways.

First, the device platform may have security features that enable the manufacturer and the network operator to exclude potentially dangerous applications from accessing bootstrapped credentials. We describe an example of this approach that uses platform security features in Nokia S60 devices in Section 5.2.

Second, many of today's mobile devices are built on top of closed platforms. In this case, access to bootstrapped credentials is restricted by default; only the platform vendor can build and install applications on it.

Finally, GAA module on an open platform device that does not have the necessary security features *must* ask the user to confirm every delivery of bootstrapped credentials to an application. In this case, the user will be (i) shown an identifier of the requesting application and the DNS name of the server, and (ii) asked to accept or reject the delivery. We think, however, that this option is not only annoying to the users but also insecure because people are unlikely to make informed decisions based on (i).

In summary, we recommend that GBA module would bootstrap as needed and provide credentials to applications silently, without involving the user. An implementation of this feature can be done on devices with platform security or closed-platform devices.

3.5.5 Split Terminal

The communication inside the terminal is not part of the 3GPP specifications, except for the communication between the smart card and the ME. The communications that take place only inside the terminal between the different kind of modules, applications and the device platform depend very much on the actual device and the corresponding implementation. Hence, the GAA specification [TS33.220] can be implemented in a split scenario, also called split-terminal scenario. In the split-terminal scenario, the 3GPP UE is split in two parts: one is the phone holding the smart card and the other part is, for example, a PC that is communicating with the phone over local link communication, e.g., Bluetooth. (See Figure 3.19.)

This split-terminal approach would enable easy login to web sites with a PC-based browser and for identity management purposes, for more details, see [Rajasekaran07]. The split-terminal scenario above applies directly to GBA_ME and 2G GBA. If the application intends to use the smart card internal GAA key in GBA_U (i.e., Ks_int_NAF) then at least one round trip for the service consumption between application in the PC and the smart card in the UE is required. The round

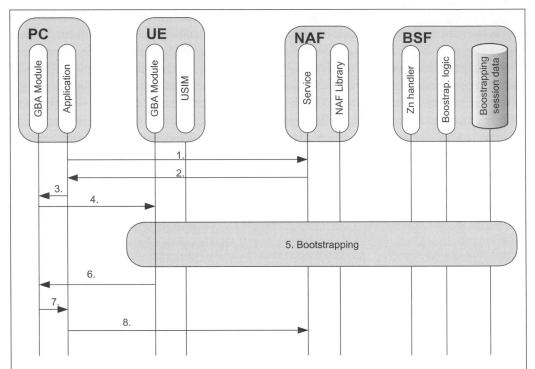

Figure 3.19.   Example message flow for split-terminal scenario

1. The application in the PC makes a request to a service-provider server that is acting as an NAF.
2. The NAF instructs the PC to use GAA to secure the communication and the normal Ua service protocol can be used (for example, as outlined in [TS24.109]).
3. The application in the PC requests now the application-specific GAA key from the local GBA Module in the PC.
4. The GBA module in the PC contacts the GBA module in the terminal and requests the application-specific GAA keys using Bluetooth or USB cable for example. The communication channel between the terminal and the PC might be secured, e.g., with Bluetooth security, usage of [TS33.259] or by other means.
5. In this step, the terminal performs the GAA Bootstrapping phase as outlined in the previous sections. The GBA module has then the NAF-specific key(s) and the related data.
6. The GBA module sends those keys to the GBA module in the PC as an answer to the secure request in step 4.
7. The GBA module in the PC forwards the NAF-specific key(s) to the application in the PC for service consumption or authentication.
8. The application in the PC can now utilize those key(s) for the service security.

trip via the GBA module is needed to hand over the encrypted data to the smart card for decryption with the Ks_int_NAF and then to return it to the PC-based application.

If there are concerns about the fact that the application-specific GAA keys are given to a PC application, then the following procedure could be used to solve the issue. If the NAF and the user side have established a common way of deriving further keys from the NAF-specific key, then either the GBA module in the terminal or in the PC can perform this task. In the service security (i.e., service consumption or authentication), the application would then receive and use these derived keys.

A variation of split-terminal scenario can also be implemented in such a way that the mobile phone uses SIM Access Profile (SAP) [SAP]. SAP enables a remote device to communicate directly with the UICC using native UICC interface commands over Bluetooth. With this approach, the GBA module functionality can be implemented in the remote device, which means that instead of the UE, the remote device will communicate with the BSF and establish the GAA master key Ks. Since most mobile phones today support SAP, the advantage of this approach is that the mobile phone does not need any additional functionalities, i.e., it can be GAA-unaware. Disadvantages are that the Ks is established in the remote device, meaning one must trust the device, and that typically when the mobile phone is in the SAP mode, its other features, like ability of receiving a phone call, are disabled. This latter disadvantage is due to the fact that the SAP is typically used with cars that are SAP-enabled, i.e., the normal phone functionality is transferred to the car phone system, and the mobile phone becomes merely just a smart card reader of the UICC communicating with car phone system over Bluetooth.

It should be noted that the above is one way to split the terminal internal functionalities for GAA. Other splits are also potentially possible, e.g., that the smart card is inserted directly in a PC using a UMTS card (or utilizing a smart card reader in the PC) and that the whole GBA module resides in the PC.

3.5.6 Interoperator GAA: Using GAA Across Operator Boundaries

The GAA functionality was described earlier in this book so that both the NAF and the UE communicate directly with the same BSF, and hence, they all function in a single MNO network. However, to take advantage of the full potential of GAA, it should be possible that NAF is associated with one MNO and the UE with another. This configuration is called interoperator GAA.

The UE always bootstraps with its home BSF. It thus always establishes the GAA master session key with its home network. However, nothing prevents the UE to contact an NAF that does not have direct association with UE's BSF. Hence, there is a need to route the requests originating from the NAF to the correct BSF. This

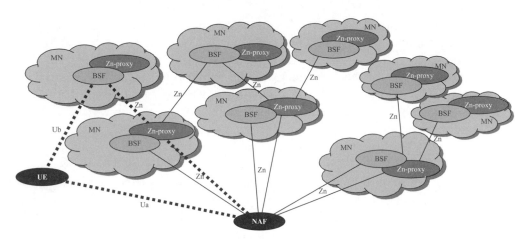

Figure 3.20.   Interoperator GAA

routing is done by an entity called Zn proxy, and it is typically operated alongside the BSF. It can also be co-located with the BSF such that they share the same Zn connection. This way the NAF does not have to distinguish between local and remote UEs. When Zn proxy receives a request from the NAF, it can route the request to the correct MNO and BSF by examining the B-TID. This is possible as B-TID contains the FQDN of the BSF. Zn proxy is also responsible to authenticate the NAF, thus, all requests forwarded by the Zn proxy are quarantined to be authentic. This way the remote BSF can make authorization decisions based on the information Zn proxy forwards. Normally, it is the BSF that authenticates the NAF, but in interoperator GAA, it is the Zn proxy. Also, as can be seen from Figure 3.20, there can be at most one Zn proxy between the NAF and the BSF, i.e., chaining of Zn proxies is not allowed.

In order to have interoperator GAA work between two MNOs, they have to:

(a) add a GAA service to their roaming agreement, and
(b) configure their GAA infrastructure to communicate with each other. In practice, this means that Zn proxies in both sides are configured to route NAF requests to other party's BSF.

This connection between the Zn proxy and the BSF is called Zn′ (Zn prime) reference point and is based on Diameter protocol. The only difference to the normal Zn reference point is the mechanism how it is protected. In the normal case, the Zn reference point is protected either by using IPsec or not protected at all. With the Zn′

reference point, the connection is protected using TLS, both client and server sides of the connection use certificates.

More details on how interoperator GAA works can be found from Annex K of [TS33.220].

3.5.7 Security Considerations of GAA

The most important security objectives for any AKA protocol are:

- Mutual authentication: each party can validate the identities of the other parties involved in the protocol.
- Key confidentiality: the resulting key must be accessible only to the parties involved in the protocol, and
- Freshness: each party can verify that the authentication is sufficiently recent and the resulting key has not been used before.

As a general-purpose AKA architecture, GAA should satisfy these objectives as well. In this section, we provide an informal discussion on the security of GAA. We start by discussing the security of the AKA protocols in the original infrastructures.

Security of Original Infrastructure

The UMTS authentication protocol is an open design that has been analyzed elsewhere and is considered to satisfy the three security objectives listed above. To recap the discussion from Chapter 2, the network is authenticated by means of the AUTN element, which also provides the freshness guarantee to the terminal. The network implicitly gets a guarantee of the freshness of authentication and session keys because RAND is an input parameter in the calculation of RES and the session keys IK and CK. The terminal is authenticated by means of the RES. Key confidentiality is achieved by generating IK and CK as functions of the long-term shared secret. Thus, the GAA bootstrapping protocol needs to make sure that these properties are preserved during bootstrapping.

The second generation (2G) cellular authentication protocols were designed to meet a lower level of security. Again, recall from Chapter 2 that authentication is only in one direction: the network does not authenticate to the terminal. As a side effect, the protocols do not provide any freshness guarantee to the terminal. The network has a freshness guarantee by virtue of RAND. Thus, the 2G variant of GAA bootstrapping protocol must compensate for the shortfalls in 2G authentication protocols by supplementing them with server authentication and freshness guarantees. Table 3.1 summarizes the security guarantees provided by the authentication protocols in the original infrastructure.

Table 3.1. Security guarantees in the original infrastructure

	Authentication	Key confidentiality	Freshness
UMTS AKA	Mutual	Yes	Yes
2G AKA	One way (terminal to server)	Yes	Server only

Both the UMTS and GSM authentication infrastructures rely on the security of those interfaces between network elements (both inter-network and intra-network interfaces) that carry authentication data or any other sensitive data. This same requirement carries over to the new interfaces introduced by GAA bootstrapping. All new interfaces can use existing network domain security mechanisms if the network elements involved are inside the operator domain. In case the NAF is operated by a third party, additional mechanisms may be needed to authenticate the NAF and secure the communication between the NAF and the BSF.

Security of Bootstrapping

GAA bootstrapping is based on HTTP Digest AKA. The particular design choice has certain implications relevant for security:

(1) HTTP Digest AKA uses MD5 as the underlying hash algorithm to compute the response for a given challenge
(2) HTTP Digest AKA uses RES as the 'password' for the Digest challenge-response calculation. RES is short (32 bits) and is sent in the clear during authentication in the original infrastructure.

If an attacker can pretend to be a base station towards a terminal, he could challenge it with a <RAND, AUTN> pair and obtain the corresponding RES. As long as AKA itself remains secure, the attacker cannot learn the AKA session keys IK and CK, and hence, the GAA master session key Ks. An attacker who learns RES in this fashion cannot therefore compromise GAA authentication as such, but can cause potential denial of service (DoS) in BSF or the terminal. DoS against BSF could be in the form of wasting BSF resources by causing it to create and store bootstrapping instances even though the genuine terminal did not request bootstrapping. DoS against the terminal could be in the form of modifying the final response from the BSF because the integrity protection of the response (Step 17 in Figure 3.6) is based on the RES. The vulnerable fields in this response are the B-TID and the key lifetime. If B-TID is modified, the terminal will store it and would detect that it is invalid only when it tries to use it in a subsequent GAA authentication over the Ua interface. Note that because of observation (2) above, the strength of the hash function is

Table 3.2.   Security guarantees in GAA bootstrapping

	Authentication	Key confidentiality	Freshness
3G GBA	Mutual (possibility of DoS exists)	Yes	Yes
2G GBA	Mutual (possibility of DoS exists)	Yes	Server: yes Client: possible

irrelevant: an attacker who learns the correct RES can calculate the correct response for HTTP Digest AKA. In other words, the use of MD5 in HTTP Digest AKA does not impact the level of security achieved by GAA bootstrapping. In summary, GAA bootstrapping preserves the original security properties of UMTS AKA but with a slight vulnerability for potential denial of service.

As we saw earlier, GAA bootstrapping in the case of 2G environment needs to compensate for the lower level of security guarantees in the original infrastructure. 2G GBA mandates that the messages exchanged during bootstrapping should make use of a server-authenticated TLS tunnel. Furthermore, the TLS server name is checked (by the terminal as well as by the BSF) against the 'realm' attribute of HTTP Digest messages. This check ensures that terminal authentication based on HTTP Digest is cryptographically bound to the server authentication based on TLS. Thus, the type of man-in-the-middle attacks identified in [ANN03] are avoided. 2G GBA does not provide any direct means to ensure freshness for the terminal. However, a terminal could locally keep track of previously seen RAND values, and thus, ensure that it does not, or at least minimize the probability that it does, use the same RAND value again. Table 3.2 summarizes the security guarantees of GAA bootstrapping.

Security of Derived Keys
Key derivation in GAA is based on HMAC-SHA256, which is believed to be a secure cryptographic message authentication function. This means that even if an attacker has discovered a number of the derived keys, he cannot find any useful information of other keys derived from the same master key. Figure 3.21 illustrates this mutual key independence aspect of key derivation.

Effect of Compromising Keys
Another point of view is to estimate the possible damage that an attacker can create by compromising each type of key.

Compromising the master key K implies that an attacker can permanently masquerade as the subscriber. The attacker can masquerade as the victim subscriber in any GAA-based service and in addition, e.g., make cellular phone calls on behalf of the victim.

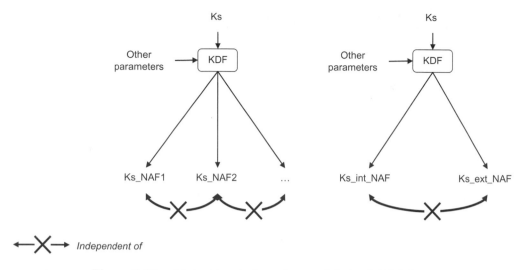

Figure 3.21. Mutual key independence of derived GAA keys

If the attack is noticed,[3] then recovery is in principle simple: the victim's subscription is terminated and a new instance of the subscription is created for the victim. In practice, a new smart card has to be given to the customer and some sort of independent authentication, e.g., physical authentication of the customer is needed.

Compromising AKA session keys CK and IK also has wider effect than GAA-based services. Until the next AKA run, i.e., as long as CK and IK are valid, an attacker can make, e.g., phone calls on behalf of the victim. In addition, as long as CK and IK, as well as keys derived from them are valid, an attacker can masquerade as the victim in any GAA-based service. Typically, new AKA run happens already before an attack is noticed, but anyway, the recovery can be done from the network side by initiating an authentication of UE. (See Figure 3.22.)

Compromising Ks is very similar to the above case, but effect of attack is restricted to GAA-based services: an attacker can masquerade as the victim in any GAA-based service. If it becomes known in BSF that a certain Ks has been compromised, then BSF can simply remove that Ks from its memory. After that, the attacker cannot

---

[3] The attack can be noticed by the victim-subscriber, e.g., when she receives the bill with accrued costs. Moreover, the MNO may notice the attack before billing the victim with the aid of fraud detection system that profiles usage patterns, such as location of the mobile device, spread of dialed numbers, call frequency and size of the bill.

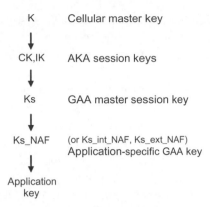

Figure 3.22.   Hierarchy of keys used in GAA-based service

derive new authentication keys for GAA-based services. The victim would need to bootstrap another Ks to authenticate to new service.

If the attacker has found some way to repeatedly compromise CK, IK or Ks, then the only permanent mitigation is the same as mentioned with compromised master key K, i.e., cancellation of subscription. If it turns out that after cancelling the subscription and changing smart card the attacker can still obtain keys, then all devices used by the subscriber need to be fixed, e.g., replaced by other devices.

The effect of compromising GAA keys Ks_NAF, Ks_ext_NAF or Ks_int_NAF is restricted to the service corresponding to the NAF in question. If it becomes known in NAF that certain key is compromised, then NAF can recover by requesting a new bootstrapping from the UE. Again, if the attacker is able to obtain Ks_NAF keys repeatedly, the only permanent cure is to cancel the subscription.

The analysis above assumed that the attacker and the subscriber are two different persons. In some cases, it is conceivable that the subscriber makes an attack against the service by utilizing compromised keys that belong to his own identity or subscription. Typically, in this kind of case, keys are used to protect application-specific data that is valuable for other users as well. For instance, Ks_NAF could be used to protect broadcast encryption key. Subscriber could try to find its own Ks_NAF and with that information discover and publish on the Internet the broadcast encryption key, thus causing losses to the broadcast service provider. In case of GBA_U, this type of attack is harder because the key Ks_int_NAF stays inside the tamper-resistant UICC.

If application keys are user-specific, then the scenario of 'user as the attacker' does not apply. Compromising the user-specific application key gives attacker an opportunity to masquerade as the victim inside this application. Requiring fresh bootstrapping would stop the attack with the compromised key. (If an application can change the compromised key without changing Ks_NAF, then new bootstrapping is not needed.)

But again, repeated compromise of user's application keys by an attacker may be a result of compromise of the higher-level keys, or in addition, it is conceivable that at the instance of the application, client has been compromised in user's device. Respectively, mitigation varies between reinstalling application and completely replacing device and/or smart card.

Another angle to the effect of compromising keys is the possibility that the attacker is able to repeat the attack against different victims. If repeated attack against different victims can be done easily, then there is vulnerability either on the application, or the device software, or device hardware, or the smart card implementation. A mitigation of this type of attack requires replacement of vulnerable part of the system, which may be costly especially in the case of hardware vulnerability.

3.6 Overview of 3GPP GAA Specifications

In this section, we want to give a brief overview of all the specifications in 3GPP that are GAA-related. GAA has grown since Release 6 to a large specification family and not all specifications are relevant for each GAA use case. We start with Figure 3.23 that illustrates the relevant specifications for implementing GAA in its basic form sorted by interfaces:

We now will introduce the rough scope of each document and the relationship to the GAA framework:

- **TS33.220 GBA** is the 'core' Generic Bootstrapping Architecture specification [TS33.220]. It describes how an application server and a terminal can be provisioned with a shared secret. There, one can find the bootstrapping procedures and their variants (including also parts of the HLR integration).
- **TR33.919 GAA** is a technical report which outlines the motivation and the relationship of GAA and some specifications [TR33.919]. It must be noted that the report is slightly outdated and gives only a limited high-level idea of the possibilities of GAA.
- **TS24.109 Ub/Ua** describes the actual bootstrapping details (Ub) using HTTP Digest AKA [RFC3310] and the framework for the applications (Ua) interface [TS24.109]. [TS24.109] describes two generic Ua protocols, i.e., GAA based on HTTP Digest and PSK TLS. Further details for the Ua protocol (i.e., the protocol using GAA) are left application-specific and can be found in the application-

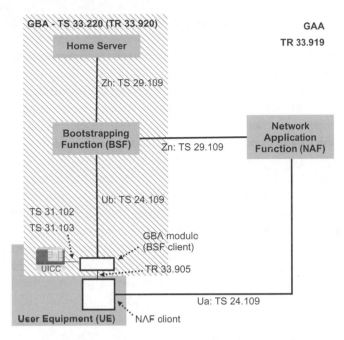

Figure 3.23.   Basic GAA specification overview

related specifications, e.g., broadcast Mobile TV. Some Ua protocols are described in Chapter 4.

- **TS29.109 Zh/Zn** describes the details of the credential fetching (Zh/Zh') and the key distribution interfaces (Zn/Zn') [TS29.109]. Zh is for the connection to the HSS and a core network interface. Zh is based on the Diameter Protocol [TS29.229] and [RFC3588], but only very limited support of Diameter is really needed. Diameter runs on top of the Stream Control Transmission Protocol (SCTP) [RFC2960].

Zh' is for the connection to an HLR and based on the MAP [TS29.002], but again only a very small part of MAP is used (for details see Chapter 5 or Section 3.3). MAP runs on top of Transaction Capabilities Application Part (TCAP) [ITU-Q.773]. TCAP itself can base either on SCTP [RFC2960] and Message Transfer Part 3 (MTP3) -User Adaptation Layer (M3UA) [RFC3332] or on the Signalling Connection Control Part (SCCP) [ITU-Q.713]. SCCP builds on top of Message Transfer Part – Level 3 (MTP3) [ITU-Q.704], MTP2 [ITU-Q.703] and the physical layer protocol MTP1.

Zn is the interface between the application server and the core network and can be either diameter-based or web service-based [W3C-SOAP], [W3C-WSDL] running on top of HTTP [RFC2616]. The Zn' is used between networks and utilizes the Diameter protocol. Chapter 5 contains many details on the implementation part of this interface (NAF library).

- **TS33.102 USIM** describes the interfaces between the phone and the USIM application on the UICC [TS33.102]. This is of special importance in the case GBA_U is used. This specification also contains the smart card details for Multimedia Broadcast Multicast Service (MBMS) security.
- **TS33.103 ISIM** describes the interface between the phone and the ISIM application on the UICC. The ISIM application can also be used for GAA.
- **TR33.905 Open Trusted Platforms** outlines some possibilities how open trusted phone platforms can realize GAA [TS33.905]. It should be noted that this technical report just gives some guidance and does not cover all potential platforms, systems or approaches that are possible in the context of [TS33.220]. Hence, it is more for giving an idea how GAA could be done in a specific scenario. For more and detailed information on how an implementation can be done on an open trusted platform see Chapter 5.
- **TR33.920 2G GBA** is an early implementation feature of Release 7 [TR33.920]. The technical report lists the change requests needed to support SIM-based bootstrapping for GAA. For the real technical information, it is better to consult the corresponding Annex in [TS33.220], where the needed details can be found. It should be noted, that some details are also in [TS29.109] and [TS24.109].

The relations between GAA specification and the protocol they use are illustrated in Figure 3.24 as a protocol stack. It should be noted, that this illustration does

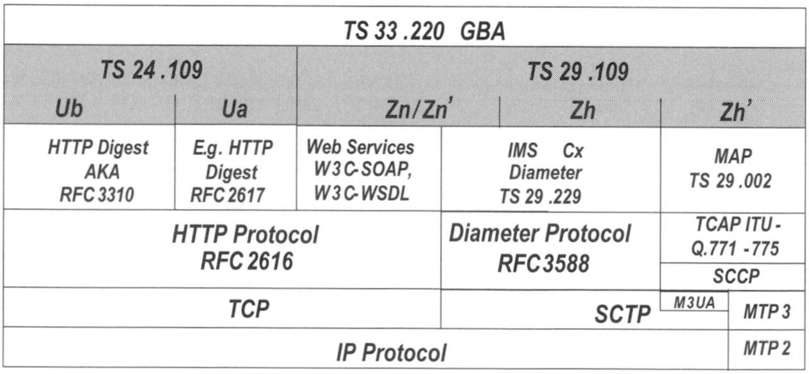

TS 33 .220 GBA				
TS 24 .109		TS 29 .109		
Ub	Ua	Zn/Zn'	Zh	Zh'
HTTP Digest AKA RFC 3310	E.g. HTTP Digest RFC 2617	Web Services W3C-SOAP, W3C-WSDL	IMS Cx Diameter TS 29 .229	MAP TS 29 .002
HTTP Protocol RFC 2616		Diameter Protocol RFC 3588		TCAP ITU - Q.771 - 775
				SCCP
TCP		SCTP	M3UA	MTP 3
IP Protocol				MTP 2

Figure 3.24.   GAA specification overview – layer view

not correspond fully to the classical Open System Interconnection (OSI) stack model.

Then we have the further enablers and 3GPP applications that build upon the GAA framework above. In those specifications below, one can find also the details for the Ua protocol depending on the actual usage of GAA:

- **TS33.221 Support for Subscriber Certificates** specification describes subscriber certificate distribution and issuing based on GAA [TS33.221].
- **TS33.222 HTTPS** specifies secure access methods to NAFs using HTTP over TLS or PSK TLS based on GAA [TS33.222]. It describes direct access to application servers using a sort of 'automated username/password' and also access to an application server through an authentication proxy.
- **TS33.259 Terminal – Remote Device Security** specification specifies a key distribution service for terminals or between a terminal and remote devices [TS33.259]. GAA as such is essentially a key generation service that provides keys between the first terminal and the network element called NAF. This NAF is acting here as a Key Centre and distributes special keys for securing the communication between the ME and the remote device (or other ME).
- **TS33.110 UICC′ – Terminal Security** specification specifies a mechanism to provision a shared key between a UICC and a terminal that may host the UICC or be connected to the device hosting the UICC via a local interface, e.g., using Bluetooth or cable [TS33.110]. It also covers, to some degree, that the terminal is not the mobile device, but, e.g., a PC. This secure channel can then be used by applications to securely communicate between the terminal and the smart card. It should be noted that TS33.259 and TS33.110 are closely aligned and can be seen as 'sister specifications' with very similar functionality, but TS 33.110 is slightly more restrictive since the smart card handles more sensitive information.
- **TR33.980 Single Sign-On (SSO)** technical report outlines the usage of GAA in combination with the Liberty Alliance Project SSO or with web services. If an operator has deployed GAA (e.g., for Mobile TV) and wants to have an SSO solution for their service providers, then [TR33.980] outlines in quite many technical details the possible approaches.
- **TS33.246 MBMS Security** specifies the usage of GAA specific keys over Ua interface and the MBMS specific key management [TS33.246]. MBMS security can use a SIM card, a USIM application on UICC or a GBA_U-aware card for bootstrapping.
- **TS33.141 Presence Security** uses GAA HTTPS [TS33.222] to secure the access to the presence service server on the Ut interface [TS33.141].

Beside those specifications from 3GPP above, there exist specifications in other standardization bodies that define GBA variants or that are using GAA, e.g., [OMABCAST].

References

[C.S0016-C]   3rd Generation Partnership Project 2 (3GPP2), C.S0016-C, *Over-The-Air Service Provisioning of Mobile Stations in Spread Spectrum Standards*, Release C, 22 October 2004. Available at http://www.3gpp2.org/public_html/specs/C.S0016-C_v1.0_041025.pdf

[C.S0023-C]   3rd Generation Partnership Project 2 (3GPP2), C.S0023, Release C, *Removable User Identity Module for Spread Spectrum Systems*, Version 1.0, April 2004. Available at http://www.3gpp.org/Public_html/specs/C.S0023-C_v1.0_060530.pdf

[ITU-Q.703]   International Telecommunication Union (ITU), *Q.703 – Signalling Link*, July 1996. Available at http://www.itu.int/rec/T-REC-Q.703/en

[ITU-Q.704]   International Telecommunication Union (ITU), *Q.704 – Signalling Network Functions and Messages*, July 1996. Available at http://www.itu.int/rec/T-REC-Q.704/en

[ITU-Q.713]   International Telecommunication Union (ITU), *Q.713 – Signalling Connection Control Part Formats and Codes*, March 2001. Available at http://www.itu.int/rec/T-REC-Q.713/en

[ITU-Q.773]   International Telecommunication Union (ITU), *Q.773 – Transaction Capabilities Formats and Encoding*, June 1997. Available at http://www.itu.int/rec/T-REC-Q.773/en

[OMABCAST]   Open Mobile Alliance (OMA), *OMA Mobile Broadcast Services*, Version 1.0 (2007). Available at http://www.openmobilealliance.org/release_program/bcast_v1_0.html

[OMASC]   Open Mobile Alliance (OMA), OMA-TS-BCAST_SvcCntProtection-V1_0–20070529-C, *Service and Content Protection for Mobile Broadcast Services Specification*, Candidate Version 1.0 – September 2007. Available at http://www.openmobilealliance.org/release_program/bcast_v1_0.html.

[Rajasekaran07]   Rajasekaran, P. Laitinen, G. Márton, R. Seidl and P. Weik, *Trust Framework and Service Delivery in SPICE*, ICIN 2007 'Emerging Web and Telecom Services: Collision or Competition?' 8–11 October 2007, Bordeaux, France.

[RFC2246]   Internet Engineering Task Force (IETF), *The TLS Protocol Version 1.0*, RFC 2246, January 1999. Available at http://www.ietf.org/rfc/rfc2246.txt

[RFC2409]   Internet Engineering Task Force (IETF), *The Internet Key Exchange (IKE)*, RFC 2409.

[RFC2616]   Internet Engineering Task Force (IETF), *Hypertext Transfer Protocol – HTTP/1.1*, RFC 2616, June 1999. Available at http://www.ietf.org/rfc/rfc2616.txt

[RFC2617]   Internet Engineering Task Force (IETF), *HTTP Authentication: Basic and Digest Access Authentication*, RFC 2617, June 1999. Available at http://www.ietf.org/rfc/rfc2617.txt

[RFC2960]   Internet Engineering Task Force (IETF), *Stream Control Transmission Protocol*, RFC 2960, October 2000. Available at http://www.ietf.org/rfc/rfc2960.txt

[RFC3310]   Internet Engineering Task Force (IETF), *Hypertext Transfer Protocol (HTTP) Digest Authentication Using Authentication and Key Agreement (AKA)*, RFC 3310, September 2002. Available at http://www.ietf.org/rfc/rfc3310.txt

[RFC3332]   Internet Engineering Task Force (IETF), *Signaling System 7 (SS7) Message Transfer Part 3 (MTP3) -User Adaptation Layer (M3UA)*, RFC 3332, September 2002. Available at http://www.ietf.org/rfc/rfc3332.txt

[RFC3546]   Internet Engineering Task Force (IETF), *TLS Extension*, RFC 3546, June 2003. Available at http://www.ietf.org/rfc/rfc3546.txt

[RFC3588]   Internet Engineering Task Force (IETF), *Diameter Base Protocol*, RFC 3588, September 2003. Available at http://www.ietf.org/rfc/rfc3588.txt

[RFC4004]   Internet Engineering Task Force (IETF), *Diameter Mobile IP v4 Application*, RFC 4004, August 2005. Available at http://www.ietf.org/rfc/rfc4004.txt

[RFC4187]   Internet Engineering Task Force (IETF), *Extensible Authentication Protocol Method for 3rd Generation Authentication and Key Agreement (EAP-AKA)*, RFC 4187, January 2006. Available at http://www.ietf.org/rfc/rfc4187.txt

[RFC4347]   Internet Engineering Task Force (IETF), *Datagram Transport Layer Security*, RFC 4347, April 2006. Available at http://www.ietf.org/rfc/rfc4347.txt

[S.S0055-A]   3rd Generation Partnership Project 2 (3GPP2), S.S0055-A, *Enhanced Cryptographic Algorithms*, Version 1.0, 15 September 2003. Available at http://www.3gpp2.org/Public_html/specs/S.S0055-A_v1.0_112403.pdf

[S.S0109-0]   3rd Generation Partnership Project 2 (3GPP2), S.S0109-0, *Generic Bootstrapping Architecture (GBA) Framework*, Version 1.0, 30th March, 2006. Available at http://www.3gpp2.org/Public_html/specs/S.S0109-0_v1.0_060331.pdf

[SAP]   Bluetooth Specification, *SIM Access Profile (SAP) Interoperability Specification*, 12 May 2005. Available at http://www.bluetooth.com/

[TCG]   Trusted Computing Group, Mobile Phone Working Group. Available at https://www.trustedcomputinggroup.org/groups/mobile

[TR33.905]   3rd Generation Partnership Project (3GPP), Technical Report TR 33.905, *Recommendations for Trusted Open Platforms*, Version 7.0.0 (2007). Available at http://www.3gpp.org/

[TR33.919]   3rd Generation Partnership Project (3GPP), Technical Report TR 33.919, *3G Security; Generic Authentication Architecture (GAA); System description*, Version 7.2.0 (2007). Available at http://www.3gpp.org/

[TR33.920]   3rd Generation Partnership Project (3GPP), Technical Report TR 33.920, *SIM Card Based Generic Bootstrapping Architecture (GBA); Early Implementation Feature*, Version 7.1.0, (2006). Available at http://www.3gpp.org/

[TR33.980]   3rd Generation Partnership Project (3GPP), Technical Report TR 33.980, *Liberty Alliance and 3GPP Security Interworking; Interworking of Liberty Alliance Identity Federation Framework (ID-FF), Identity Web Services Framework (ID-WSF) and Generic Authentication Architecture (GAA)*, Version 7.4.0 (2007). Available at http://www.3gpp.org/

[TS23.003]   3rd Generation Partnership Project (3GPP), Technical Specification TS 23.003, *Numbering, Addressing and Identification*, Version 7.4.0 (2007). Available at http://www.3gpp.org/

[TS24.109]   3rd Generation Partnership Project (3GPP), Technical Specification TS 24.109, *Bootstrapping interface (Ub) and Network Application Function Interface (Ua); Protocol Details*, Version 7.5.0 (2006). Available at http://www.3gpp.org/

[TS29.002]   3rd Generation Partnership Project (3GPP), Technical Specification TS 29.002, *Mobile Application Part (MAP) specification*, Version 7.9.0 (2007). Available at http://www.3gpp.org/

[TS29.109]   3rd Generation Partnership Project (3GPP), Technical Specification TS 29.109, *Generic Authentication Architecture (GAA); Zh and Zn Interfaces based on the Diameter protocol; Stage 3*, Version 7.5.0 (2006). Available at http://www.3gpp. org/

[TS29.229]   3rd Generation Partnership Project (3GPP), Technical Specification TS 29.229, *Cx and Dx interfaces based on the Diameter protocol; Protocol details*, Version 7.6.0 (2007). Available at http://www.3gpp.org/

[TS31.102]   3rd Generation Partnership Project (3GPP), Technical Specification TS 31.102, *Characteristics of the Universal Subscriber Identity Module (USIM) Application*, Version 7.8.0, (2007). Available at http://www.3gpp.org/

[TS31.103]   3rd Generation Partnership Project (3GPP), Technical Specification TS 31.103, *Characteristics of the IP Multimedia Services Identity Module (ISIM) Application*, Version 7.1.0 (2006). Available at http://www.3gpp.org/

[TS33.110]   3rd Generation Partnership Project (3GPP), Technical Specification TS 33.110, *Key Establishment between a UICC and a Terminal*, Version 7.3.0 (2007). Available at http://www.3gpp.org/

[TS33.141]   3rd Generation Partnership Project (3GPP), Technical Specification TS 33.141, *Presence Service, Security*, Version 7.1.0 (2006). Available at http://www.3gpp.org/

[TS33.220]   3rd Generation Partnership Project (3GPP), Technical Specification TS 33.220, *Generic Authentication Architecture (GAA); Generic Bootstrapping Architecture*, Version 7.7.0, (2007). Available at http://www.3gpp.org/

[TS33.220]   3rd Generation Partnership Project (3GPP), Technical Specification TS 33.220, *Generic Authentication Architecture (GAA); Generic Bootstrapping Architecture*, Version 7.7.0 (2007). Available at http://www.3gpp.org/

[TS33.221]   3rd Generation Partnership Project (3GPP), Technical Specification TS 33.221, *Generic Authentication Architecture (GAA); Support for Subscriber Certificates*, Version. 6.3.0 (2006). Available at http://www.3gpp.org/

[TS33.222]   3rd Generation Partnership Project (3GPP), Technical Specification TS 33.222, *Generic Authentication Architecture (GAA); Access to Network Application Functions Using Hypertext Transfer Protocol over Transport Layer Security (HTTPS)*, Version 7.2.0 (2006). Available at http://www.3gpp.org/

[TS33.223]   3rd Generation Partnership Project (3GPP), Technical Specification TS 33.223, *Generic Authentication Architecture (GAA); Generic Bootstrapping Architecture (GBA) Push function*, version 0.1.0 (2007), Release 8. Available at http://www.3gpp. org/

[TS33.246]   3rd Generation Partnership Project (3GPP), Technical Specification TS 33.246, *3G Security; Security of Multimedia Broadcast / Multicast Services*, Version 7.3.0 (2007). Available at http://www.3gpp.org/

[TS33.259]   3rd Generation Partnership Project (3GPP), Technical Specification TS 33.259, *Key Establishment between a UICC Hosting Device and a Remote Device*, Version 7.0.0 (2007). Available at http://www.3gpp.org/

[TS51.011]   3rd Generation Partnership Project (3GPP), Technical Specification TS 51.011, *Specification of the Subscriber Identity Module – Mobile Equipment (SIM-ME) interface*, Version 4.15.0 (2005). Available at http://www.3gpp.org/

[W3C-SOAP]   World Wide Web Consortium (W3C): *Simple Object Access Protocol (SOAP)*, (2007). Available at http://www.w3.org/TR/soap/

[W3C-WSDL]   World Wide Web Consortium (W3C): *Web Services Description Language (WSDL)*, Version 2.0 Part 0: Primer, (2005). Available at http://www.w3.org/TR/2005/ WD-wsdl20-primer-20050803/

[X.S0004-540-E] 3rd Generation Partnership Project 2 (3GPP2), X.S0004-540-E, *MAP Operations Signaling Protocols*, Version 1.0.0, March 2004. Available at http://www.3gpp2.org/ Public_html/specs/X.S0004-540-E_v1.0_040406.pdf

[Yan04]   Jeff Jianxin Yan, Alan F. Blackwell, Ross J. Anderson, Alasdair Grant: *Password Memorability and Security: Empirical Results*. IEEE Security & Privacy 2(5): 25–31 (2004).

4

Applications Using Generic Authentication Architecture

4.1 Standardized Usage Scenarios

In this section, the usage scenarios of the GAA that are standardized at the time of this book being written are described. GAA functionality can also be used for many other purposes, five of which we will outline in the following section.

4.1.1 Authentication Using GAA

The most straightforward way of utilizing GAA for authentication is to use the established ***GAA credentials as username and password***, namely the Bootstrapping Transaction Identifier (B-TID) and the NAF-specific key (Ks_NAF). Most application protocols include the possibility to authenticate entities with username/passwords, thus using GAA in those protocols is simple in principle. In practice, the complexity actually comes from the fact that also other kinds of credentials exist. Thus, there is the need to negotiate and agree which credentials are to be used among the parties. For this purpose, 3GPP has specified two generic profiles [TS33.222], [TS24.109]:

Cellular Authentication for Mobile and Internet Services
Silke Holtmanns, Valtteri Niemi, Philip Ginzboorg, Pekka Laitinen and N. Asokan
© 2008 John Wiley & Sons, Ltd

- **GAA usage with HTTP Digest**, which profiles how GAA-based authentication is used in HTTP-based protocols.
- **GAA usage with PSK TLS**, which profiles how TLS can use GAA to establish a secure connection.

These two generic profiles offer GAA support automatically to a wide range of protocols that run on top of the Transport Control Protocol (TCP) [RFC0793] and the User Datagram Protocol (UDP) [RFC0768]. With HTTP Digest profile, the GAA-based authentication can be used in all web browser-based services provided that the browser supports GAA. With PSK TLS profile, the GAA-based authentication support is added to all TCP- and UDP-based protocols as all those protocols can be tunnelled through TLS. Thus, these profiles can be used in majority of the existing protocol used in Internet today.

GPP has also specified a mechanism to **bootstrap a Public Key Infrastructure (PKI) utilizing GAA** [TS33.221]. One of the toughest challenges with PKI is how to do the initial registration of an entity to the system, i.e., how a Certification Authority (CA) or a Registration Authority (RA) is able to do the authentication the very first time. This happens typically when user requests a certificate to be issued, i.e., to be enroled to a PKI system. For example, this can be done by the end-user physically entering a registration office and presenting his ID like driver's license or passport. Or it can be bootstrapped from another system like GAA where user has already done some kind of registration (in GAA case with the MNO), by, e.g., presenting his ID and giving his billing address. After the **subscriber certificate** based on GAA has been issued, it can be used in any PKI-based system that accepts the corresponding root certificate or CA certificate from the CA that issued it.

GAA can also be used in configurations where the application service logic and the authentication function are separated into two different entities. This can be done in several ways, which can be roughly divided in two categories:

- **Proxy mode**, where a proxy is set between the user terminal and application server that does the authentication of the user on behalf of the application server.
- **Referrer mode**, where an authentication server authenticates the user and communicates with the application server directly or indirectly.

An example of proxy mode configuration is a Wireless Application Protocol (WAP) gateway, where incoming HTTP requests from a mobile phone are always routed through a proxy that can determine the user's identity based on the IP address and add the identity information to the HTTP requests. An example of referrer mode configuration is a Liberty Alliance Identity Federation Framework (ID-FF) 1.x or the

OASIS Security Assertion Markup Language (SAML) 2.0, where a dedicated server called Identity Provider (IdP) authenticates users based on some mechanism and then asserts to application logic servers called Service Providers (SP) that the end-user has been authenticated with appropriate method and provides some SP-specific identity information to it.

Both 3GPP and the OMA have specified how GAA is used in the proxy mode. There, the proxy first authenticates the user-based HTTP Digest profile, and then upon successful authentication, forwards the request to an application logic server that trusts the proxy.

3GPP has also written a technical report on how GAA can interwork with Liberty Alliance ID-FF, and the Web Service Framework (ID-WSF) as well as OASIS SAML 2.0 in the referrer mode [TR33.980]. There an IdP uses HTTP Digest or PSK TLS profile to authenticate the user and then asserts to the SP that the user has been authenticated using GAA. This same architecture can be used in any identity management system such as Microsoft's CardSpace and community-developed OpenID where the IdP or Security Token Service (STS) uses GAA to authenticate to user and then interacts with the SP or Relying Party (RP) to assert that the user has been authenticated.

4.1.1.1 HTTP Digest Authentication

The usage of GAA in HTTP Digest [RFC2617] is described in [TS33.222] and [TS24.109]. Normal HTTP Digest as it is currently implemented in most web browser clients use username and password to calculate hash response to the challenge that the server sends to the client. Typically, end-user is prompted with a dialog by the browser where end-user is supposed to type in the username and password. With GAA, this username and password are replaced by GAA's B-TID as username and NAF-specific key as password.

The GAA usage in HTTP Digest needs to be negotiated between the client and the server. The client may indicate to the server that it is GAA-aware by including a special product token to the User-Agent HTTP header that the client sends to the server. This token contains either '3gpp-gba' or '3gpp-gba-uicc'. In the former case, the client is able to use the NAF-specific key in the Mobile Equipment (ME) (i.e., Ks_NAF or Ks_ext_NAF). In the later case, the client is also able to use the NAF-specific key available in the UICC smart card (i.e., Ks_int_NAF) and is only applicable to the GBA_U (see Chapter 3 for more details on the GBA variants). This indication is optional according to the specification, but helps the server to decide whether GAA should be used or not.

The server in turn must indicate that it expects the client to use GAA credentials and this is done using the 'realm' parameter in the HTTP Digest challenge. The realm attribute is set to either '3GPP-bootstrapping@<naf-dns-name>' or

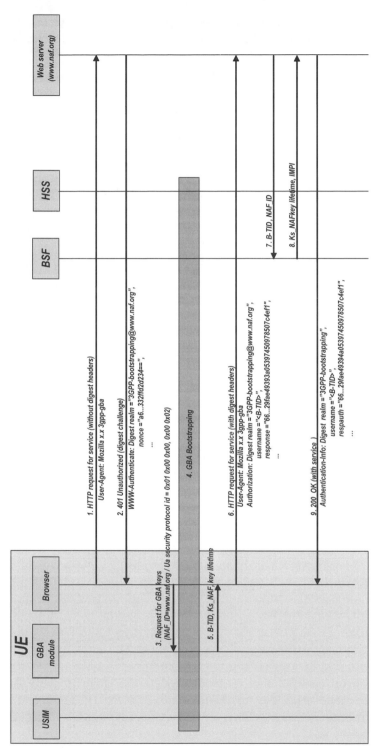

Figure 4.1. GAA used with HTTP digest

'3GPP-bootstrapping-uicc@<naf-dns-name>'. When the client application receives such a challenge, it knows that GAA should be used; either it is ME-based (i.e., GBA_ME / 2G GBA) as in the former case, or UICC-based (i.e., GBA_U) as in the latter case. The client should also use server's Domain Name Service (DNS) name that is indicated in the realm attribute when deriving the NAF-specific key, but it should also check that the DNS name in the realm is the same as in the client's original HTTP request.

In Figure 4.1, there is an example message flow showing how GAA is used with HTTP Digest. The terminal contains a web browser that is GAA-enabled and it contacts a web server that is GAA-aware. The protocol used here is HTTP-based but it could use, for example, the SIP [RFC3261] as well.

The message sequence below can be run both in the plain mode without TLS or inside a TLS tunnel (a.k.a. HTTPS) where the TLS session has been established based on server certificate. In the former case, the Digest authentication's Quality Of Protection (QOP) should be set to 'auth-int' in order to provide HTTP payload message integrity as well. In the latter case, the integrity protection, as well as confidentiality, is provided by the underlying TLS tunnel.

1. Web browser is instructed (e.g., user types in an URL using either http:// or https:// protocol schema) to a web site called www.naf.org. For each outgoing HTTP request, the web browser adds a product token '3gpp-gba' to indicate that it is GAA-aware and HTTP Digest with GAA is supported.

   If https:// protocol schema was used, then the HTTP request is sent through a TLS tunnel that is established beforehand.
2. Upon receiving the HTTP request, the web server notices that the URL in the request identifies a service (or resource) that requires authentication. In addition, the web server inspects the User-Agent header of the request, and discovers that the GAA is supported by the browser. The web browser is configured to use GAA-based authentication for this particular service if the web browser has indicated that GAA is supported. Thus, it starts to use HTTP Digest to authenticate the client.

   The web browser sends an HTTP response with status code 401 'Unauthorized' indicating that authentication is required. Furthermore, it adds a WWW-Authenticate header to the response containing the normal HTTP Digest parameters with a special 'realm' value: '3GPP-bootstrapping@www.naf.org'. The web server uses the prefix value '3GPP-bootstrapping@' to trigger the web browser to use GAA, and the postfix value 'www.naf.org' to indicate to the browser what should be the NAF's Fully Qualified Domain Name (FQDN) of the NAF_ID.

   When the web browser receives the 401 'Unauthorized' request, it first examines the 'realm' value whether the Digest authentication should be based on GAA. If the prefix value of the 'realm' is not '3GPP-boostrapping@', then the web browser would prompt the end-user for username and password normally. Since the prefix of the 'realm' does instruct the browser to use GAA, it will continue with GAA-based authentication.

3. The browser extracts the postfix value of the 'realm', which is the FQDN of the NAF that should be used to construct the first part of the NAF_ID. But first it must check that the FQDN from the realm is the same in the FQDN of the URL that was used in step 1. If this check is successful, the browser constructs the whole NAF_ID by concatenating the FQDN of the NAF with a five-octet long Ua security protocol identifier, which in HTTP Digest case is 0x 01 00 00 00 02. Then the browser sends a request for GAA key to the GBA module of the terminal using an API and passing the NAF_ID with, optionally, other parameters to the GBA module.

4. The GBA module performs bootstrapping involving the security module (e.g., USIM application on the smart card), the BSF and the HSS to establish a GAA master key (Ks). If GBA module has already the GAA master key that is valid, it can use the existing Ks directly without contacting the other entities.

   Typically the GBA module will also check whether the application requesting NAF-specific key is authorized to do so. This can be done in several levels, e.g., to check whether the application is, in general, authorized to access GBA module and request any key using any NAF_ID (coarse-grained access control). Also, the GBA module could check whether the application is authorized to request a particular key using a particular NAF_ID (fine-grained access control). This is discussed more in Section 5.2.

5. The GBA module derives the NAF-specific key using the NAF_ID provided by the web browser and other parameters that have been stored to the GBA module like IP Multimedia Private Identity (IMPI) corresponding to the security module (e.g., USIM), and RAND of the bootstrapping session. It then returns the B-TID, the NAF-specific key, the key lifetime, and optionally, also the type of the GAA bootstrapping that was done (e.g., was the bootstrapping based on 2G GBA or GBA_U).

6. Upon receiving the GAA material from the GBA module, the web browser calculates the HTTP Digest response to the challenge. In Digest calculations, it uses the B-TID as the username and the NAF-specific key as the password, and then normally calculates the response according to [RFC2617]. It will then resend the original HTTP request (same as in step 1) to web server containing the Digest response parameters in the 'Authorization' header.

7. When the web server receives the HTTP response, it extracts the B-TID for the 'username' parameter and uses it and the NAF_ID to request the NAF-specific key from the BSF. In this simple example, only the ME-based key is requested and no special indications like requests for service-specific USSs using GSIDs or GBA_U awareness flag are included.

8. The BSF checks first that the NAF in question is authorized to use the particular NAF_ID. If it is authorized, the BSF locates the GAA master key identified by the B-TID and calculates the NAF-specific key, similarly as what the GBA module did during step 4.

   In this example, the BSF has been configured to return also the IMPI to the NAF, so that it is able to map the user with a persistent record in its local databases. The other way to do this is to use USSs to send service-specific user identities to the NAF. This would be used if the NAF requires a persistent identity of the user that is different from the IMPI. Third option is not to send any persistent identity and in this case, the NAF would just know that the user is a valid subscriber of the corresponding operator but would not know exactly who the subscriber is. The last case is good for situations where a service is not user-specific but it is intended for customers of certain operator in identical manner.

9. The web browser validates the Digest response to the challenge using the B-TID and the NAF-specific key. If this validation is successful, the user is authenticated. Then the web browser can authorize the user to use the service based on its local policies.

   Alternatively, the USS can be used to indicate to the NAF whether the user is authorized to use a service or a part of a service. This aspect is discussed in more detail in Section 3.3.

   If the user is authorized, the web server will return the service requested by the browser. The web server also includes the 'Authentication-Info' header, which the browser validates using the B-TID and NAF-specific key as specified in [RFC2617].

Any subsequent HTTP requests to the web server are also authorized using the 'Authorization' header as was done in step 6 and the same goes for the corresponding HTTP responses with 'Authentication-Info' header. The receiving entity must always validate these Digest headers.

4.1.1.2 Pre-Shared Key TLS

The usage of GAA in TLS is described in [TS33.222] and [TS24.109]. It is based on Pre-Shared Key (PSK) TLS that is specified in [RFC4279], where a TLS session is established based on a Pre-Shared Key identity (PSK-identity) and PSK. Similar to HTTP Digest, GAA credentials can also be used to establish the TLS session by simply setting PSK-identity to be the B-TID and PSK itself to be the NAF-specific key.

As in HTTP Digest case, the usage of GAA in PSK TLS needs to be negotiated between the client and the server and this happens during TLS handshake. The client can indicate (indirectly) that it supports GAA-based authentication by including one or more PSK-based ciphersuites in the ClientHello message. This message must also contain the server_name TLS extension [RFC3546] to indicate the DNS name of the NAF. The server will respond with standard ServerHello and ServerKeyExchange messages, where the ServerKeyExchange will contain psk_identity_hint parameter either '3GPP-bootstrapping', '3GPP-bootstrapping-uicc', or both for indicating the User Equipment (UE) what GAA keys are acceptable to the NAF. The UE will then use either the ME or UICC-based key depending on whether the application resides on the ME or on the UICC, respectively.

In Figure 4.2, there is an example message flow how GAA is used with PSK TLS. The terminal contains a service-specific application that is GAA-enabled, and it contacts a server that is GAA-aware.

TLS session has now been established between the client application and the server. The client application and the server can run any application-specific protocol inside the TLS tunnel.

4.1.1.3 Proxy Mode Authentication

The general architecture in the proxy mode can be seen in Figure 4.3. In this architecture, the authentication function has been separated from the service logic. Typically, the authentication proxy (AP) that is handling authentication, and optionally, also authorization of the user, is functioning in front of the UE. For example, if the protocol used between the UE and the service is based on HTTP, then the AP can be a reverse HTTP proxy, where the set-up, if configured, is in such a way that the DNS names (like server1.operator.com) are mapped to the IP address of the AP. Thus, when the UE is contacting server1.operator.com, it actually connects to the AP. The AP then authenticates and potentially authorizes the user and the forwards the request to the correct application server. The application servers are typically only accessible through the AP, thus only authenticated access is possible.

The proxy mode authentication has been specified in both 3GPP and in OMA. In 3GPP, the authentication proxy functionality has been specified in [TS33.222], where the protocol between the UE, the Authentication Proxy (AP) and application servers is based on HTTP. In 3GPP, it is also possible to do authorization in the AP using functionalities provided by the USS. In OMA, the proxy mode functionality is called aggregation proxy and it has been specified in the context of XML Document Management (XDM) [OMAXDM], where the aggregation proxy can optionally use GAA to do the authentication. However, in XDM, the aggregation proxy does only authentication and no authorization. Instead, the latter is done always by the application server (for more details see Section 4.1.3).

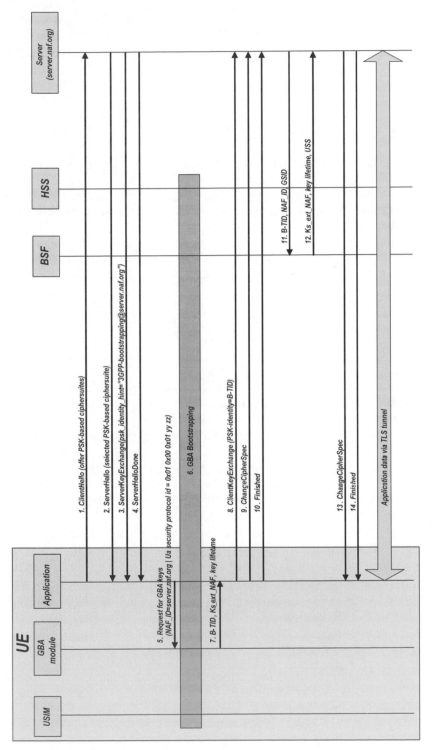

Figure 4.2. GAA used with PSK TLS

Figure 4.2.   GAA used with PSK TLS (*continued*)

1 The application in the UE starts communicating with the server by initiating the TLS handshake messaging using the ClientHello message. As an indication of GAA awareness, the client includes one or more PSK-based ciphersuites to the ClientHello message. The ClientHello also includes the server_name extension to indicate to the server what FQDN it is using to contact the server. This is particularly important if multiple FQDNs (e.g., virtual hosting) can be used to contact the server. In this case the server_name is 'server.naf.org'.

2 The server receives the ClientHello message and notices that PSK-based ciphersuites are present in the message. It has been configured so that if client includes any PSK-based ciphersuites to the ClientHello message, it will use GAA for setting up the TLS session. It sends first ServerHello message containing one of the PSK-based ciphersuites included in the ClientHello message.

3 The server will then send the ServerKeyExchange that will include the psk_identity_hint value. As the server is configured to communicate only with an application residing in the ME (and not in the UICC) the psk_identity_hint value will contain only '3GPP-bootstrapping'. This will indicate to the client application that only ME-based keys are acceptable to the server.

4 The server sends ServerHelloDone message to indicate to the client that it is waiting for next TLS handshake messages from the client application.

5 The application determines when it received the ServerHello and ServerKeyExchange message that GAA- and ME-based key should be used to set up the TLS session. The client application constructs the NAF_ID by setting the prefix to be 'server.naf.org' and the postfix, i.e., the Ua security protocol identifier to be 0x 01 00 01 *yy zz* where the *yy* and *zz* are the PSK-based ciphersuite codes defined in TLS 1.0.

   The application will then contact the GBA module and request GAA key using the NAF_ID.

6 The GBA module performs bootstrapping involving the security module (e.g., USIM), the BSF and the HSS to establish a GAA master key (Ks). If GBA module has already the GAA master key that is valid, it can use the existing Ks directly without contacting the other entities.

   Typically, the GBA module will also check whether the application requesting NAF-specific key is authorized to do so. This can be done in several level, e.g., to check whether the application is, in general, authorized to access GBA module and request any key using any NAF_ID (coarse-grained access control), or also in addition the GBA module could check whether the application is authorized to request a particular key using a particular NAF_ID (fine-grained access control). Access control is discussed more in Section 5.2.

7 The GBA module derives the NAF-specific key using the NAF_ID provided by the application and other parameters that have been stored to the GBA module like IMPI of the security module (e.g., USIM), and RAND of the bootstrapping session. It then returns the B-TID, the NAF-specific key, the key lifetime, and optionally, also the type of the GAA bootstrapping that was done (e.g., Was it based on 2G GBA or GBA_U?).

Figure 4.2. GAA used with PSK TLS (*continued*)

8 The application sends the ClientKeyExchange message to the server containing the psk_
identity value, which in this case is the '3GPP-bootstrapping'; concatenated by the
B-TID.

9 The application sends ChangeCipherSpec message to indicate to the server that cipher
should be changed.

   The application derives the TLS premaster secret from the NAF-specific
key.

10 The application sends Finished message that contains the hash of the handshake
messages.

11 When the server receives the messages in steps 8–10, it extracts the B-TID from
the ClientKeyExchange, and uses it and the NAF_ID to request the NAF-specific
key from the BSF. In this simple example, only the ME-based key is requested, and
thus, no GBA_U-aware indication is sent. Also, the server has been configured to
request a User Security Settings (USS) from the BSF using a GAA Service Identifier
(GSID).

12 The BSF checks first that the NAF in question is authorized to use the particular NAF_ID.
If it is authorized, the BSF locates the GAA master key identified by the B-TID and
calculates the NAF-specific key the same way as the GBA module did during step 4. It
also tries to locate the USS requested by the NAF using the received GSID value. If the
USS is found from subscriber's GUSS, the corresponding USS value is returned to the
NAF (if it is allowed by the BSF local policy).

   In this example, the BSF has been configured not to return the IMPI to the NAF, thus,
IMPI is not sent to the NAF. The mapping of the user to a persistent record in NAF's
local databases is done using USS and the persistent identity inside it. This mechanism
is typically used if the NAF requires a persistent identity other than IMPI.

   The server validates the Finished message received from the application in step 10.

13 The server sends ChangeCipherSpec to indicate to the server that cipher should be
changed.

   The server derives the TLS premaster secret from the NAF-specific key.

14 The server sends Finished message that contains the hash of the handshake messages.

   Once the application receives the Finished message, it validates it.

As example of the AP functionality, we take the 3GPP authentication proxy
scenario, where the AP does both the authentication and authorization. (See
Figure 4.4.) The details of how GAA-based authentication is done with HTTP
Digest are left out from this example flow as they have been described already
in Section 4.1.1.1. This sequence flow concentrates on the AP-specific
functionalities.

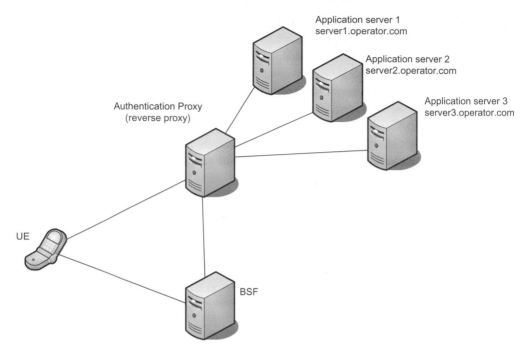

Figure 4.3. Proxy Mode Authentication

One simple configuration option for the authentication proxy should be noted and that is the possibility to use the AP in front of a web server providing a service that is intended to one operator's subscribers. In that case, no X-3GPP-* headers would be used and no identity information is transferred from the BSF to the AP: The AP would just make sure that the application is used by a valid subscriber of an MNO, but there would not be any user-specific services hosted on the server.

4.1.1.4 Referrer Mode Authentication

The general architecture in the referrer mode authentication can be seen in Figure 4.5. It resembles well-known identity management systems like Liberty Alliance and Web Services Secure Exchange (WS-SX), OpenID. In all those systems, the authentication mechanism itself is not specified and any acceptable authentication mechanisms could be used. GAA can nicely fit in to these architectures by providing the authentication towards the Identity Provider (IdP). The IdP would function as a NAF

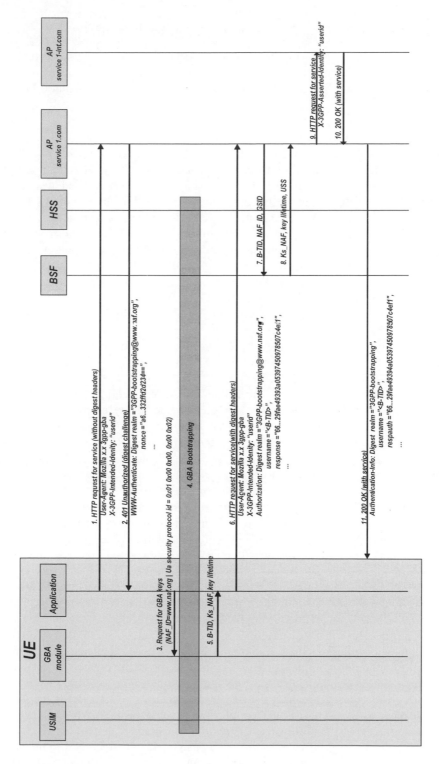

Figure 4.4.   Authentication proxy usage with GAA – message flow

Figure 4.4.   Authentication proxy usage with GAA – message flow (*continued*)

  1 An application in the UE requests a service. It opens a connection to a 'server1.com'
     and sends an HTTP request to the service. The application will also send an indication
     to the service what is the identity of the user, and this is sent in 'X-3GPP-Intended-
     Identity'.
  2 The network has been configured in such a way that the DNS entry for 'server1.com'
     is assigned to the IP address of the AP. The AP examines the request and notices that
     it is directed to 'server1.com' by inspecting the HTTP headers of the request. The AP
     has been configured in such a way that GAA-based authentication and authorization
     is required for this particular service, thus, it sends a 401 Unauthorized message with
     WWW-Authentication header back to the client (as was done in Section 4.1.1.1).
3–5 These steps are exactly the same as the steps 3–5 described in Section 4.1.1.1.
  6 The application resends the same request to the service as it did in step 1, but now it
     includes the Authorization header that contains the HTTP Digest response (calculated
     as described in Section 4.1.1.1).
  7 Upon receiving the request, the AP determines that it must request GAA keys and a
     certain service-specific USS identified by a GSID from the BSF. It requests the GAA
     key and the USS by sending the B-TID, the NAF_ID and the GSID to the BSF.
  8 The BSF authorizes the AP and checks that it authorized to use the NAF_ID in question,
     and to receive the USS identified by the GSID. If the AP is authorized, the BSF
     proceeds by calculating the GAA key Ks_NAF, and returns it, the key lifetime and the
     USS (if it was found from subscriber's GUSS) to the AP.
  9 The AP first validates the Digest part of the request received in step 6 using the Ks_
     NAF. If the validation succeeds, the AP then examines the received USS if such was
     received. The USS contains one identity and the AP checks that this identity is the
     same as in the received X-3GPP-Intended-Identity header. If the check is successful,
     the AP forwards the request to the actual server that is handling the application logic.
     The AP removes the Authorization header from the request as it already processed it
     and the application server does not need to process it further.
        The AP removes the X-3GPP-Intended-Identity header and adds the X-3GPP-
     Asserted-Identity header to the outgoing request to indicate to the application server
     which was the identity that has been authenticated by the AP. The application server
     can use this information to check whether the identity or identities in the HTTP payload
     match this information.
        In this case, the authorization in the AP is done by merely checking that the USS
     was present in subscriber's GUSS. If the USS would not have been in the GUSS, no
     USS would have been returned from the BSF to the AP, and the AP would have taken
     this as an indication that the subscriber in question is authorized to use the server.
 10 The application processes the incoming request based on its application logic. If there
     is identity information in the service-specific elements of the protocol (e.g., in HTTP
     payload), the server checks that those identities are the same as what was given to it
     in the X-3GPP-Asserted-Identity header. Then, it returns the response to the AP.
 11 The AP receives the response from the application server, adds the WWW-Authentication
     header to the request, and forwards the request to the application in the UE. The UE
     processes the response. If necessary, the UE may make a new request to the service,
     in which case, the steps from 6 to 11 are executed again with different payloads.

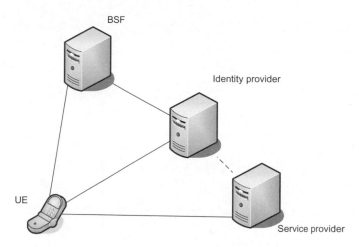

BSF

Identity provider

UE

Service provider

Figure 4.5. Referrer mode authentication

towards the BSF and use the identity management protocols of the system towards the service provider. In this way, the identity management system and GAA are orthogonal and complement each other.

As an example, we can take SAML 2.0-based Web Browser SSO functionality that happens between a web server ran by a service provider, GAA-enabled web browser in the UE, and IdP and BSF operated by an MNO. (See Figure 4.6.) This scenario, together with other possibilities related to Liberty Alliance Project and SAML specifications, is also described in [TS33.980].

The SP also creates a session with the browser in such a way that all subsequent requests from the browser can be mapped to this authenticated session. This can be done by using, for example, HTTP Cookies or URL-based session keys.

If the terminal had already a valid GAA session, then the bootstrapping with the BSF in step 6 would not be done but the existing GAA master key would be used. Also, if the browser (i.e., end-user) had already been authenticated by the IdP before the browser contacts our SP, then the IdP would have had an authenticated session with the browser before it contacted the SP. In this case, the HTTP request in step 3 would already contain the Authorization header as in step 8 and the IdP could just issue a new assertion to our SP. Thus, only steps 1, 2, 3, 8, and 11–15 would have been executed.

4.1.2 Broadcast Mobile TV Service

Broadcast mobile television enables people to view broadcasted programs on their mobile phones. They search and select those programs through a service guide that is also broadcasted by the service provider. (The main advantage of broadcast over

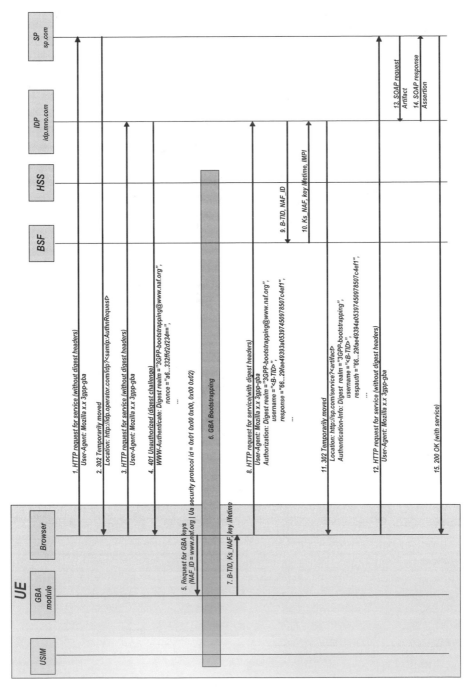

Figure 4.6. Referrer mode authentication usage with GAA – message flow

Figure 4.6. Referrer mode authentication usage with GAA – message flow (*continued*)

1 The user instructs the browser to contact a service provider (SP hosting an sp.com service by either using a bookmark or by typing in the URL. The browser sends the HTTP request the SP and the request also includes the '3gpp-gba' indication in the User-Agent header.

2 When the SP receives the request, it determines that for using the service, the end-user must be authenticated using SAML 2.0. Thus, it obtains the identity provider and sends a redirect HTTP response with the URL to the selected IdP to the browser. The URL also contains the to the browser. The URL of the IdP is obtain for example using preconfigured IdP or the end-user can select the IdP from a list of IdPs. In the latter case, there would be additional HTTP response and request messages between the UE and the SP to do the selection.

3 The browser reacts normally to the redirect request and contacts the IdP using the URL given by the SP. The URL contains the encoded. As always, the browser also included the '3gpp-gba' to the User-Agent header, and thus, the IdP can decide to authenticate the end-user using GAA.

4 Upon receiving the request from the browser containing the and the IdP has been configured that GAA-based authentication is used. Thus, it sends a 401 Unauthorized message with WWW-Authentication header back to the client (as was done in Section 4.1.1.1).

5–7 These steps are exactly the same as the Steps 3–5 described in Section 4.1.1.1.

8 The browser resends the same request to the IdP as it did in step 1 but now it includes the Authorization header that contains the HTTP Digest response (calculated as described in Section 4.1.1.1).

9 Upon receiving the request, the IdP requests the GAA key from the BSF using the received B-TID. In this example case, the IdP has not been configured to use USS so it does not include any GSIDs. In turn, the BSF has been configured to return the IMPI of the user to the IdP.

10 The BSF authorizes that the IdP is allowed to use the particular NAF_ID and derives the GAA key. It returns the GAA key to the IdP together with the IMPI.

11 The IdP verifies that the Digest response received from the browser in step 8 was correct using the GAA key. If the check is successful, the IdP uses the IMPI to locate the user profile data from its local databases. It also generates an authentication assertion that contains the SP-specific information about the authenticated user and an artefact that is used later by the SP. Then the IdP sends a redirect message to the browser with a URL to the SP that contains the SAML artefact.

12 The browser reacts normally to the redirect request and contacts the SP using the URL received from the IdP in step 11.

13 Upon receiving the request, the SP extracts artefact from the URL and uses it to fetch the authentication assertion from the IdP using Simple Object Access protocol-based (SOAP-based)[1] back channel.

14 The IdP locates the authentication assertion from its databases using the artefact and returns it to the SP.

15 The SP examines and verifies the received assertion, and if this check is successful, provides the requested service to the browser.

---

[1] http://en.wikipedia.org/wiki/SOAP

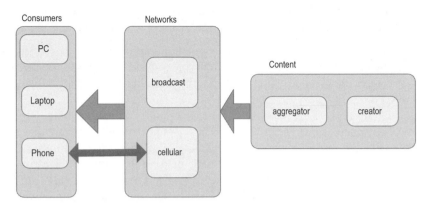

Figure 4.7.   Mobile television service entities

unicast mobile TV is the economy of network resources achieved by broadcasting.) Figure 4.7 depicts the main entities that cooperate in providing this service: creator of content, aggregator of content, MNO and broadcast network operator. Content is typically created by media companies, aggregated into channels and delivered to consumers over broadcast network. Through the cellular network, consumers subscribe to the service and choose the programs to view.

Broadcast mobile television is a new service and there are different ways to organize the business around it. Of particular interest to MNO could be the case where it maintains service ownership in terms of branding, customer relationship and billing relationship. In this business model, it is to the advantage of the MNO to build essential parts of the service on cellular infrastructure. This potential advantage is one of the main reasons why OMA BCAST smart card profile was standardized.

Mobile television and Multimedia Broadcast Multicast Service (MBMS), in which encrypted content is wirelessly broadcast or multicast, are the lead applications driving the deployment of GAA. In those applications, the delivery of service keys that are needed to decrypt the received content is secured with GAA. This makes valid cellular subscription a prerequisite for, e.g., watching mobile television programs. Also, when generating charges for mobile television and MBMS, the person's cellular identity, which is verified with GAA, is used when generating charges for mobile television and MBMS.

In broadcast mobile television, standardized by Open Mobile Alliance (OMA) BCAST group, the content is broadcast using Digital Video Broadcasting – Handheld (DVB-H) technology [DVB-H]. In MBMS, the content is multicast over cellular radio network. For instance, the content may be targeted to handsets in a particular radio cell, or a group of cells. But, since the security of OMA BCAST smart card profile[2] is based on MBMS specifications [TS33.246], both systems have almost identical security architecture providing data protection, key distribution and authentication. Moreover, OMA BCAST lists MBMS as one of the possible broadcast bearers, in addition to DVB-H. This enables a deployment where OMA BCAST specifications define the application layer and MBMS the underlying broadcast layer.

4.1.2.1 Security Goals

The main security goal in protecting the mobile television service is that only those having both (i) valid cellular subscription and (ii) valid service subscription can view protected programs broadcasted using DVB-H or MBMS technology.

That goal is implemented as follows. Broadcasted programs are encrypted with traffic keys. Traffic keys, which change often, are encrypted with service key and broadcasted as well. The service key is specific to a service or program; it is obtained by each terminal individually via a point-to-point IP connection, such that:

(1) connection establishment requires a terminal containing a working cellular smart card with USIM or SIM application, and
(2) the validity of user's subscription to mobile television service is verified before the service key is delivered to the terminal.

Thus, the user must have a valid (i) cellular subscription and (ii) mobile television service subscription, to obtain the service key.

4.1.2.2 Service Architecture

An example of system providing mobile television service that uses OMA BCAST smart card profile security is depicted in Figure 4.8. The main network interfaces are marked with their labels in the BCAST Architecture document [OMASA] and the GAA specification [TS33.220].

The content is encoded, encrypted and streamed[3] by Content provider (interface SP-5-1). The content is encrypted with Traffic Encryption Key (TEK) that is changed frequently by the content provider.

---

[2] OMA BCAST group specification [OMASC] contains two ways (profiles) to secure the broadcasted content: Digital Rights Management (DRM) profile and Smart Card profile. The DRM profile is using PKI and digital rights management (DRM) technologies. In this section, we concentrate on the smart card profile that is based on shared keys and GAA.
[3] The streaming protocol is Real–time Transport protocol [RFC3550] and the encryption method is either Secure Real-Time Transport Protocol (RTP) profile [RFC3711], or ISMACryp [ISMACryp].

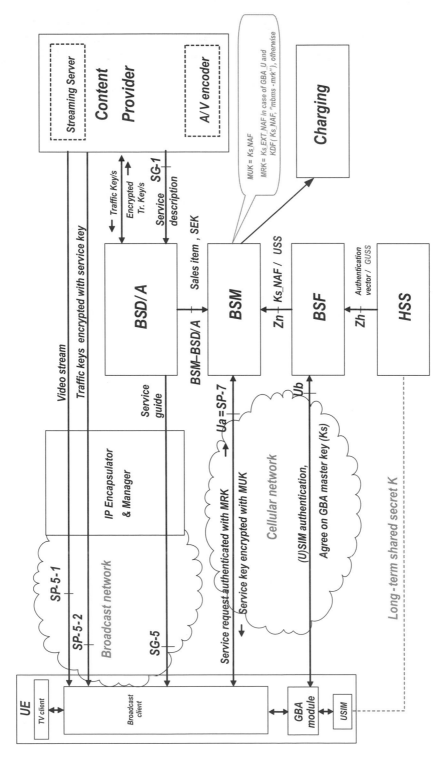

Figure 4.8. Mobile television service with OMA BCAST smart card profile security

DVB-H network broadcasts data in MPEG-2[4] content streams. To save receivers battery power, the DVB-H broadcast of each channel is not continuous; it occurs in short bursts, e.g., once per two seconds. This enables a receiver to reduce its power consumption while unwanted channels are being broadcast. Also, a channel may be intended to be viewed only in part of the regions covered by the DVB-H network. For those reasons, the system contains an IP Encapsulator and Manager, which encapsulates IP packets into MPEG-2 transport streams, buffers the content between the transmission bursts and routes the content to the intended regional transmitters.

Before the transmission starts, the BCAST Service Distribution / Adaptation (BSD/A) server receives a service or program description from the content provider (interface SG-1). BSD/A then does three things: First, it generates a Service, or Program Encryption Key (SEK/PEK). That key typically stays the same for the duration of the program. During the broadcast, the content provider periodically delivers one or more short-lived TEKs to BSD/A, which returns them back encrypted with SEK/PEK. The content provider streams the encrypted traffic keys with the broadcasted content (interface SP-5-2). Thus, users that have SEK/PEK can obtain traffic keys from the broadcasted data and decrypt the content. The sequence of broadcasted traffic keys encrypted with SEK/PEK are encapsulated in a stream of, so called, Short Term Key Messages (STKM). Each STKM contains a TEK.

Second, BSD/A includes the service description and the SEK/PEK identifier in the broadcasted guide to the currently available services (interface SG-5).

Third, BSD/A delivers the sales item, i.e., the identifier of the service or program, together with SEK/PEK to the BCAST Subscription Management (BSM) server.

The BSM server contains a database of subscribers to the mobile television service. It handles delivery of SEKs/PEKs to paying subscribers and generation of charging records based on those deliveries. Those records will be sent either immediately or in batches to a charging function. The BSM server verifies cellular subscription and obtains mobile subscriber identity from the cellular network. Therefore, it includes a NAF module and Zn interface to BSF.

The User Equipment (UE) contains a cellular smart card with a long-term cellular subscriber key K; a copy of K is kept in the HSS. The UE also contains two applications that are specific to mobile television: a broadcast client that receives and decrypts broadcasted data and a TV client that plays the decrypted content on audio–visual hardware of the device. In addition, UE contains a GBA module that has interfaces with the broadcast client, the USIM application on cellular smart card UICC and an external interface Ub with the BSF.

The communication between UE and BSM server is over a point-to-point IP connection. Two keys are derived from the bootstrapped master session key Ks to secure that communication: MBMS Request Key (MRK) and MBMS User Key (MUK).

---

[4] http://en.wikipedia.org/wiki/MPEG-2

When the user that has been viewing the broadcasted services guide chooses a program, the broadcast client in the UE sends a service request to BSM server over Ua interface (SP-7 in OMA BCAST architecture [OMASA]). If the service request has been successfully authenticated with MRK and the user is allowed to view the requested program, then the BSM server will send the SEK/PEK encrypted with MUK to the UE. The message containing SEK/PEK encrypted with MUK is called Long-Term Key Message (LTKM).

A detailed sequence of messages that is triggered by user receiving a broadcasted programs guide and requesting one of those programs is shown in Figure 4.8. In essence, the sequence implements authenticated request for a broadcasted program and reply with traffic keys that encrypt that program. It may seem laborious to implement a single function with 32 messages (the 28 messages described below, together with three messages between BSD/A and content provider and one message from BSD/A to BSM, which are not shown in Figure 4.8). Recall, however, that we deal with a distributed system with eight components: USIM, GBA module, Broadcast client, BSF, HSS, BSD/A, BAM and Content Provider. There are 32/8 or four messages on the average per component, i.e., two request/reply message exchanges on the average per component.

4.1.2.3 Message Flow Example

Figure 4.9 shows the sequence of events from the moment a broadcasted Service Guide is received in the UE broadcast client (message 0) until the UE broadcast client receives one or more traffic encryption keys for decrypting the broadcasted data from the GBA module (message 27). Square brackets '[ ]' indicate message fields that are optional in the standard and '*' indicates possible repetition of a field.

In that example flow, we have made the following main assumptions:

i. The database of subscribers to the mobile television service in the BSM server has been initialized and subscriber's cellular identity, IP Multimedia Private Identity (IMPI), is used to retrieve subscription records in that database.
ii. The UE contains a GBA_U-aware smart card. This fact is recorded in the subscriber's record in HSS. Also, the UICC has an application for receiving and storing the SEK/PEK key. This application is called MBMS key Generation and Validation Storage and Function (MGV-S/F).

We make the first assumption because we think that this is how the mobile television service is going to be deployed initially. We make the second assumption to illustrate the use of GBA_U.

In Chapter 2, we described service-independent messages that comprise the bootstrapping of master session key Ks from the long-term cellular subscriber key K

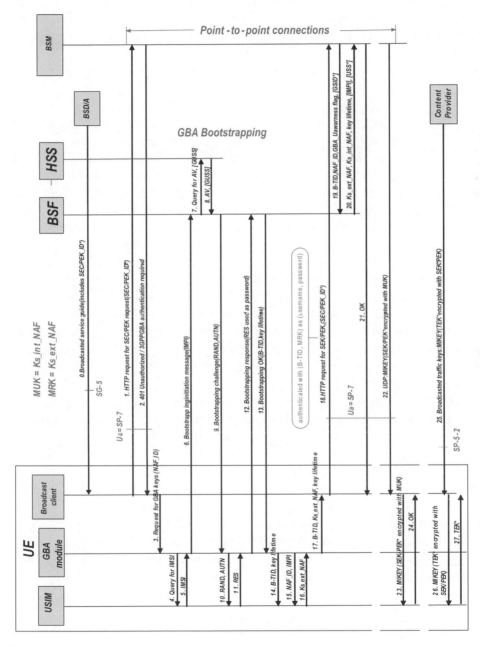

Figure 4.9. Message sequence in OMA BCAST smart card profile with GBA_U (GBA-aware Release 6 USIM).

(messages 6 to 13 in Figure 4.9) and the GAA Zn interface (messages 19 and 20). Network messages that are specific to the mobile broadcast service can be grouped as follows: message 0 contains the broadcasted service guide; messages 1, 2, 21 and 22 comprise the Ua interface; and message 25 contains the broadcasted Traffic Encryption Keys (TEKs) that are themselves encrypted with the service or program key (SEK/PEK).

We conclude this section with detailed description of message sequence in Figure 4.9.

0. The broadcast client in the UE receives a service guide that is broadcasted over DVB-H radio by BSD/A. Among other things, the guide includes a list of program or service descriptions and the identifiers of the SEK/PEK keys that are needed to decrypt the corresponding program or service.[5]

1. When the user chooses a service from the guide, the broadcast client in the UE initiates an HTTP connection with the BSM server associated with the service. Over that connection, it sends an HTTP request containing all the SEK/PEK_IDs that identify the encryption keys of the desired service or program. This request does not yet contain any security-related data.

2. The BSM server responds with the 401 'unauthorized' challenge message. That message instructs the broadcast client to use GAA for authentication: The response includes the WWW-Authenticate header with HTTP Digest fields. The realm field in that header contains the Domain Name Server (DNS) name of the BSM server. To illustrate, if 'BSM.com' is the DNS name of the BSM server, then the value of the realm field would be '3GPP-Bootstrapping@BSM.com'.

3. After receiving the challenge message from the BSM server, the broadcast client detects that GAA credentials should be used to authenticate the SEK/PEK request. Therefore, it requests GAA credentials from UE GBA module. This request contains at least the NAF_ID with BSM server's DNS name and the Ua interface security protocol identifier.[6]

4–14. If the UE GBA module does not have a valid GAA master key (Ks) and B-TID, it initiates the bootstrapping procedure with the BSF; otherwise, it will proceed to message 15. In this example, the UICC is GAA-aware. This fact is recorded within the GUSS in HSS. Therefore, HSS adds GUSS to message 8 and the BSF-specific information element in GUSS contains a 'GBA_U awareness' flag.

---

[5] The SEK/PEK identifier is a 7-byte long string in the 'ProtectionKeyID' element that can be attached to, e.g., 'Service' or 'Content' fragment of the service guide.
[6] The Ua protocol identifier is a string of five octets. Its value is 0x 01 00 00 00 01 in MBMS and OMA BCAST smart card profile cases.

15. In this example, the USIM application in the UICC is GBA-aware and so the NAF-specific key (Ks_ext_NAF) need to be derived on the UICC. Thus, the GBA module sends the NAF_ID and the IMPI to the USIM application.

16. The USIM application derives the two NAF-specific keys: Ks_ext_NAF, and Ks_int_NAF, based on its internal master key (Ks) and the input parameters. Ks_int_NAF is stored on the UICC; Ks_ext_NAF is returned to the GBA module.

17. GBA module responds to the broadcast client request (step 3) by returning the NAF-specific key (Ks_ext_NAF), B-TID and key lifetime.

18. The broadcast client calculates the correct response to the challenge message (step 2) using HTTP Digest method with B-TID as username and Ks_ext_NAF as password. It then sends a new HTTP request for SEK/PEK to the BSM server. The calculated response is included in the Authorization header of that message.

19. The BSM server receives the request and extracts the B-TID from the request. It then uses the B-TID to request the GAA keys from the BSF. Included to this request are also the NAF_ID, GBA_U awareness flag, and optionally, some GAA Service Identifier (GSID) identifying USS entries. The GBA_U awareness flag indicates to the BSF that BSM server is capable to use both Ks_int_NAF and Ks_ext_NAF keys.

20. The BSF locates the correct GAA session from its local databases using the received B-TID and calculates the two bootstrapped keys Ks_int_NAF and Ks_ext_NAF. It then returns these keys alongside with the key lifetime and requested User Security Settings (USSs) provided that they exist. The BSF will also send the IMPI of the USIM to the BSM server.

21. Upon receiving the information from the BSF, the broadcast server validates the request coming from the broadcast client (step 18) using the Ks_ext_NAF. If this validation is successful, then the broadcast server uses the received IMPI to locate the mobile TV subscriber record in its database and updates that record with data received from BSF in message 20. It will then send 200 'OK' message to the broadcast client indicating a successful authentication. Upon receiving that message, the broadcast client starts to monitor incoming traffic on the User Datagram Protocol (UDP) port 2269.

22. The BSM checks if the user is authorized to receive the requested service or program, e.g., by comparing the age of the user to the intended age of the program's audience. If the check is successful, then the server creates a charging record for the service usage and sends the requested SEK/PEK to the broadcast client over UDP in a Multimedia Internet Keying (MIKEY) [RFC3830] protocol message.

23. Upon receiving the MIKEY message, the broadcast client forwards it to the MGV-S/F application on the UICC for further processing. The broadcast client continues to monitor the traffic on the UDP port 2269 to receive possible updates of SEK/PEK in the future.

24. The MGV-S/F application decrypts the received SEK and PEK keys from the MIKEY message using Ks_int_NAF, stores them locally in the UICC and returns OK message to the broadcast client to indicate successful SEK/PEK key delivery. The decrypted SEK/PEK never leaves the smart card.

25. The steps 25–27 happen when the content server broadcasts one or more new TEK keys. This may happen even quite often, e.g., once in 20 seconds. The content server broadcasts the new TEK keys encrypted with SEK/PEK in a MIKEY protocol message over UDP.

26. Upon receiving the MIKEY message, the broadcast client forwards it to the MGV-S/F application on the UICC for further processing.

27. The MGV-S/F application decrypts TEK using SEK/PEK it has received in step 24 and returns TEK to the broadcast client. The broadcast client uses TEK to decrypt the broadcasted stream.

4.1.2.4 Tracing Source of Leaked Keys

In summary, broadcast mobile TV is a service that distributes service-specific keys (SEK/PEK) to subscribers that are authenticated with GAA: the BSM server acts as NAF.

Since the service-specific keys are shared between many subscribers, this service is susceptible to 'user as an attacker' scenario that we have mentioned in Section 3.5.6: The attacker could obtain the SEK by tampering his own terminal or smart card and anonymously publishing that SEK on the Internet. (We assume that it is impractical for the attacker to publish the traffic encryption keys because they change very frequently.) Once SEK is revealed, anybody can unscramble the program it protects without paying.

If this attack happens, then the broadcast service provider can use the fact that each subscriber receives his SEK over an authenticated point-to-point connection with the BSM server to trace a subscriber, or a group of subscribers, that are leaking the SEK.

For example, if there are four subscriber groups: G1, G2, G3 and G4, and one of those groups is always leaking the received SEK, then the guilty group can be identified in two iterations by encrypting the same TEKs stream with two different SEKs in each iteration:

- First iteration: Subscriber groups G1 and G2 will receive SEK1; subscriber groups G3 and G4 will receive SEK2.

Suppose the broadcast service provider observes that SEK2 is revealed, then it learns that the guilty group is either G3 or G4 and proceeds to the second iteration.

- Second iteration: For the next broadcast, subscriber groups G1, G2 and G3 receive SEK1' and subscriber group G4 receives SEK2'.

    If SEK1' is revealed, then service provider learns that the guilty group is G3. If SEK2' is revealed, then the operator learns that the guilty group is G4.

The above example describes the basic idea of the so called 'traitor tracing' schemes: Traitor tracing in this case is a game between the broadcast service provider and a clique of 'traitors', i.e., subscribers that repeatedly publish a broadcast service encryption key, thus causing monetary loss to the service provider. (Members of the traitor clique may secretly communicate with each other when deciding on what SEKs to reveal.)

To trace traitors, the service provider encrypts the same broadcasted TEK stream with several different SEKs and divides the subscribers into groups. Each group receives a different SEK. The service provider then observes which of those keys are revealed and in the next iteration uses another set of SEKs and divides the subscribers again to narrow the set of potential traitors.

An important parameter in this game is the number of TEK streams that are broadcasted simultaneously for the same program or service. On the one hand, the tracing is quicker with more TEK streams. For instance, in the above example, the traitor group could be revealed in a single iteration if the service provider could encrypt the same TEKs stream with four different SEKs. On the other hand, more TEK streams mean less bandwidth for the broadcasted service itself.

In the DVB-H case, the broadcast of a single TEK stream uses about 100 B/s. The broadcast of a single DVB-H program or service needs about 400 KB/s. Therefore, assuming that the broadcasted content bandwidth can be reduced by up to 0.5% without noticeable degradation in its reception quality, there should be enough bandwidth to simultaneously broadcast up to 20 TEK streams, each encrypted with a different SEK.

For more information on traitor tracing, we refer the reader to the Ph.D. thesis of T. Martin [Martin04] and the references therein.

4.1.3 Further Standardized Usage Scenarios

In this section, we briefly outline further standardized usages of the GAA, mainly the ones used outside of 3GPP. The Open Mobile Alliance [OMA] defines network-agnostic standards for a large range of services. OMA works closely aligned with 3GPP on services that utilize GAA. We already mentioned that the OMA Broadcast Group defined the smart card profile for content and service protection in broadcast scenarios, like broadcast Mobile TV. OMA also uses GAA for location security and XML Document Management. An enabler is a central concept in OMA. It is a function or specification that can be used by many applications and other specifications.

The OMA security group has defined a GBA Common Security Function Enabler [OMAGBAProf]. The GBA Profile outlines a profile of GBA as defined by 3GPP and 3GPP2. There exist several options in the GAA specifications and the OMA Enabler Specifications tries to reduce the number of potential interoperability scenarios and potential problem cases. The idea is that the GBA Profile is then used throughout all OMA Enabler Specifications which want to utilize GAA. OMA GBA Profile supports SIM, USIM, ISIM and UIM.

OMA XML Document Management (XDM) is used, for example, by OMA Presence and Availability Working Group (PAG) in their OMA Presence SIMPLE specification (see Section 5.5.4 in [OMAPres]). The XDM specification is using the XML Configuration Access Protocol (XCAP) [RFC4825] and enables manipulation of individual XML elements and, in particular, attributes instead of the whole XML document. The terminal management server and terminal use HTTP Digest for peer-to-peer authentication and confidentiality to the aggregation proxy. The role of the aggregation proxy is to act as a front-end for the XDM servers and take care of the confidentiality and authentication to the XDM clients, i.e., the mobile terminals. Hence, the terminals access the Resource List XDM server that store the watchers presence list through the aggregation proxy. The authentication in OMA XDM specification [OMAXDM] is based on [TS33.222], or alternatively, on using HTTP digest authentication, which also could make use of GAA. (See Figure 4.10.)

The 3GPP Presence Service Security [TS33.141] is using GAA on the Ut interface between the terminal and the presence server (or respectively, the authentication proxy acting as a GAA NAF). The terminal and the presence server (respectively the authentication proxy) support TLS as outlined in [TS33.222]. Analogous to the OMA Presence [OMAPres], the terminal authenticates either directly to the presence server or respectively to the authentication proxy using [TS 33.222], if 3GPP methods are used. The usage of shared secret provided by [TS33.220] is also envisioned. The server side is authenticated using HTTPS as outlined in [TS33.222]. As user identities, the presence server may use IMPI or IMPUs obtained through the Zn interface from the BSF server. The terminal sends it preferred public identity and then the presence server or the authentication proxy checks the validity (further details can be found in [TS33.141]).

The OMA Secure User Plane Location (SUPL) service [OMASUPL] uses GAA in their SUPL security functions. OMA SUPL supports use cases, where a terminal roams into another network and the position is obtained with support from home and visited network for the network resident mobile location service. PSK TLS can be used, as outlined in [TS33.222], for mutual authentication between the SUPL-enabled

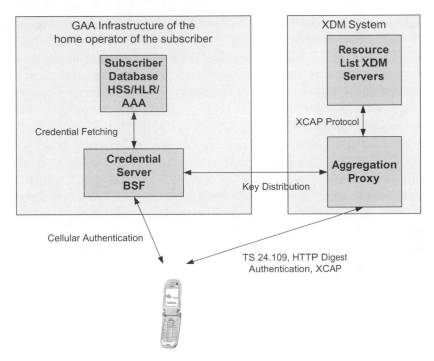

Figure 4.10. XML document management using GAA. Reproduced by permission of ©
Nokia

terminal and the SUPL location centre server that is acting as a GAA NAF. In 3GPP2
deployments, the corresponding 3GPP2 specifications can be used.

Recently also, standardization bodies dealing with different kind of fixed and cable
networks started to integrate the GAA into their specifications as a security enabler
for their services. The fixed-mobile convergence is an ongoing activity that is not
yet finalized (by the time of writing). We will outline the current status of integration
as far as it is at least, to some extent, stable work in progress.

The European Telecommunications Standards Institute (ETSI) has a core compe-
tence centre for fixed networks and for migration from circuit-switched networks to
packet-based networks. The Telecommunications and Internet converged Services
and Protocols for Advanced Networking [TISPAN] standardization body, whose
work include the convergence aspects for xDSL and PSTN/ISDN, uses GAA as an
option in the user authentication on the Ut interface. The Ut interface is used in the
IP Multimedia Subsystem (IMS) between the terminal and the application server.
Over the Ut interface, the terminal can manipulate own data (for instance, authorization policies, public service identities, etc.) of IMS-based SIP services, like messaging, presence or conferencing. The Ut interface is protected using HTTP over TLS
and can either be based on [TS33.222] or HTTP Digest authentication. Even though

HTTP digest seem to have a lower priority in the specification work than, for example, ISIM-based authentication, it is probable that due to the large existing legacy, the HTTP Digest authentication will play an important role. In the Ut interface case, the application server can also be accessed through an authentication proxy. (See Figure 4.11.)

The usage of GAA for TISPAN has been discussed also for other scenarios, like media security or the split-terminal scenario, but due to the big time pressure for the Release 1 of TISPAN, no further steps were taken at that point of time to work out the details. At the time of writing, some interest appeared in 3GPP Security group to study in more detail the optimization possibilities for the split-terminal scenario, where the GBA module resides not in the mobile phone, but in a PC. This is ongoing work and the outcome is not yet clear. In principle, the split-terminal scenario can be done within the framework of [TS33.220].

Another consortium that started to have a closer look at GAA is CableLabs. The members of CableLabs are from the packet cable television industry with North American focus. The existing 3GPP IMS specifications have been extended (in Release 8 of 3GPP) to accommodate their specific needs, to enable cable opera-

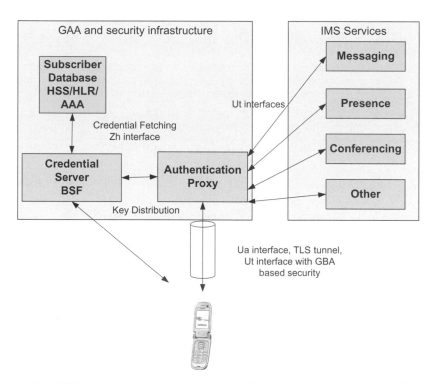

Figure 4.11. IMS service authentication using GAA. Reproduced by permission of © Nokia

tors to use the new telecommunications technologies and be interoperable with the existing mobile infrastructure. CableLabs have also created extensions to the GAA specification TS33.220 [CableLabsGBA] and TS29.109 [CableLabsZh] with a new method to bootstrap the service-specific credentials. In these cases, the GAA specification has been extended with a new bootstrapping mechanism called GBA_H.

GBA_H extends the HTTP Digest authentication over TLS to GAA. The bootstrapping procedure starts with the establishment of an HTTPS tunnel between the device (note that this does not need to be a mobile phone) and the BSF. HTTP Digest authentication mechanism is then used to establish the needed credentials and derive the session keys between the network and the device. This implies that the BSF would need some additional functionality and also the device needs to support this new GAA variant. At the time of writing, the GAA extensions of CableLabs have not yet been transferred to and embraced by 3GPP; they are CableLabs-specific and may change when/if brought into 3GPP for common standardization.

4.2 Additional Usage Scenarios

In this section, we sketch some additional potential scenarios where GAA may be used to solve the authentication needs of applications. These use cases have not yet been standardized. The reasons vary. Standardization is an expensive process involving many players with different business interests and background. Also, standardization bodies focus on a special area and do not consider use cases that fall outside their area.

Nevertheless, whenever there is a compelling usage scenario, *de facto* solutions addressing the usage scenario often emerge prior to formal standardization. This section is intended to inspire such solutions by illustrating the potential of deploying GAA in five example scenarios. The descriptions in this section should therefore be taken as illustrations rather than as detailed implementation specifications.

We begin by sketching two considerations of applying GAA to a new scenario. First, GAA has an embedded trust assumption: all parties involved in GAA trust MNOs. Before applying GAA to a specific scenario, a designer should check if that assumption holds for the parties involved in that scenario. Second, the original cellular infrastructure was designed for cellular services. GAA is therefore applicable for protecting assets that are similar in value to cellular services. While it is difficult to put exact monetary figures for the value of cellular services, one can spot scenarios where the assets are clearly outside the range for which the cellular authentication infrastructure is intended. For example, while it may be reasonable to use GAA to pay for a meal, it is certainly not reasonable to use it to pay for a car or a house.

We now describe five example scenarios and discuss how GAA may be applied to them.

4.2.1 Secure Enterprise Login

Many companies protect the access to their corporate network with one-time password, i.e., a password that is valid for a short time, e.g., one minute. Typically, the employee receives a hardware token, such as SecurID™ produced by the RSA company. The token generates and displays one-time passwords, which the employee types into an application in his terminal, e.g., a browser or a VPN (Virtual Private Network) client. The client establishes a secure connection with Internet Key Exchange (IKE) protocol or HTTP-based authentication schemes such as HTML forms and HTTP Digest. The corporate access server generates the same sequence of one-time passwords and so can verify the employee's access. To protect against access by outsiders who obtained lost or stolen hardware tokens, the employee must enter another, long-term password during the authentication.

This authentication method is more secure than entering the long-term password directly into the VPN client, or into a browser, because it requires both the knowledge of that password and the possession of the company-issued hardware token.

Its disadvantages include:

i. the costs of the tokens' logistics; and
ii. the inconvenience of the employees who must remember to carry the tokens.

Both disadvantages can be alleviated by using the employee's mobile phone instead of special-purpose hardware token. An employee will typically have a mobile phone close by while working: The UICC within the mobile phone being the 'hardware token' for accessing the corporate network can reduce (i) because no additional hardware needs to be provisioned[7] and it can reduce (ii) because people are used to carry their mobile phones. In the rest of this section we will describe how this can be implemented with GAA.

The main idea of this implementation is that MNO will help the corporate remote access server to verify that the client has a certain (U)SIM and the corporate server will verify that the user knows the right password P.

Care must be taken to prevent MNO learning that password from publicly exchanged messages. (Otherwise an attacker that, e.g., penetrates the MNO's system will be able to access the corporate network.) This is done by the employee's terminal and the corporate server both deriving a new key Ks_co from Ks_NAF and from P,

---

[7] Software will need to be provisioned to the mobile phone, but corporate applications can be installed today on mobile phones by the IT administrators and employees.

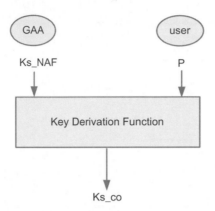

Figure 4.12.   Derivation of corporate access key Ks_co

as shown in Figure 4.12. That key will be used for mutual authentication of remote access. In this scheme, the corporate access server acts as NAF and it must receive B-TID before mutual authentication takes place. The corporate access server will receive the identity of the employee which it needs to choose the right P from the BSF.

If the smart card in the employee's terminal is GBA_U-enabled, then Ks_co will be derived from Ks_ext_NAF instead of the Ks_NAF.

The password P could be identical to the employee's corporate network password, in which case, the corporate remote access server needs an interface to the database where intranet passwords are kept.

The derivation of Ks_co may use more inputs in addition to Ks_NAF and P. For instance, Figure 4.13 shows derivation of Ks_co from a device-specific key D, in addition to Ks_NAF and P. This configuration ensures that (i) the Ks_co comes from a certain device, which (ii) has the right (U)SIM and that (iii) the employee knows P.

As an example, the employee's terminal and the corporate server could agree on the key D when the terminal is inside the company premises,[8] after which D is bound to employee's identifier (e.g., to the MSISDN of the employee's mobile phone). The security of that agreement is based on both parties being in secure corporate network.

---

[8] The key agreement could be initiated, e.g., when the employee touches an RFID tag at the company's service point with his terminal.

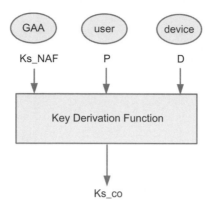

Figure 4.13. Derivation of corporate access key Ks_co with additional input D

4.2.2 *Personalization for Payments and Securing Public Transport Tickets*

The promise of a mobile device as the personal trusted device is a popular and recurring idea. Such a personal trusted device will help the user securely access a wide variety of services. Many of these services already have existing authentication and access control mechanisms. For example, banks already use different kinds of one-time password authentication mechanisms; enterprises use physical access control tokens to allow building access; public transit companies use similar tokens for ticketing. Therefore, to fully realize the potential of a mobile trusted device, the mobile device must be able to support different types of existing and new authentication mechanisms. Already, there are different approaches to embedding secure environments in mobile devices [OBCPoster; OBCTR; Venyon]. These are general-purpose secure environments built into the phones during manufacture. Usually, there is a mechanism to validate a device as having a compliant secure environment. One way to enable this device validation is to initialize a unique device key pair in the device during manufacture and have the manufacturer issue a certificate for the device public key.

To use such a device in a particular application, it has to be *personalized*: the correct cryptographic keys and other information for a given user have to be securely provisioned to the secure environment on the user's device. While the device validation mechanisms (like device public key and certificate) provide a way to verify that the target device does indeed have a valid secure environment, a provisioner who needs to send credentials to a user needs a way to identify the *specific* device of that particular user.

Figure 4.14 shows one way in which GAA can be used to solve this personalization problem. We assume that a user has a UE with a secure environment and would

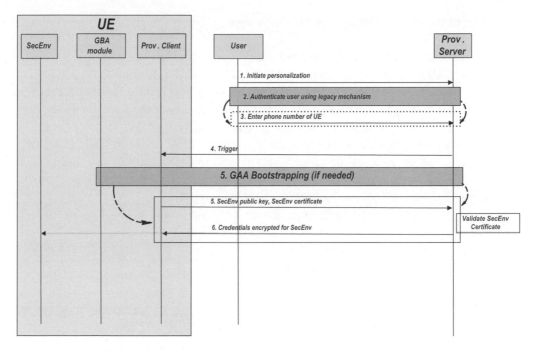

Figure 4.14.   Personalization of secure tokens using GAA

like to provision credentials for some application like online banking. The secure environment is identified by a device public key and an associated device certificate. The provisioning server has some legacy mechanism to authenticate the user. For example, the user may have a username/password he already uses for online banking.

The user goes to the provisioning server to initiate provisioning by accessing an HTTP URL (step 1). The server authenticates the user using the legacy authentication mechanism (step 2). For example, this may be done by the user visiting a TLS-protected web site of the server, and providing his legacy username and password. This results in a secure channel between the user and the provisioning server. The user can now type in the phone number of the UE to which the credentials should be provisioned (step 3). Since this is done via the secure channel, the phone number is securely bound to the initial user authentication.

The provisioning server now contacts a provisioning client on the UE by sending it a trigger (e.g., a binary SMS message). The UE and the Provisioning Server now perform GAA bootstrapping, with the Provisioning Server acting as an NAF. This results in a secure channel between the provisioning server and the provisioning client on the UE. GAA guarantees the provisioning server that the end point of this channel corresponds to the correct cellular identity, namely the phone number entered by the user in step 3.

Now the provisioning client can send the device public key and the device certificate of the UE secure environment (step 5). Since this information is sent via a GAA-bootstrapped secure channel, the provisioning server is guaranteed that the device public key in fact corresponds to the device to which the user wants the credentials provisioned. The provisioning server can verify that the device public key in fact corresponds to a legitimate secure environment by verifying the device certificate. It can then encrypt the credentials using the device public key and send it back to the provisioning client who will pass them on to the secure environment where the credentials can be decrypted and used (step 6).

This is one way of personalizing a device for a particular type of secure token. The attractiveness of using GAA is the ease of user interactions: for example, steps 1 and 2 can be done in several other ways such as having the user call a service desk number and provide the phone number of his UE.

4.2.3 Secure Messaging in Delay and Disruption-prone Environments

Over the last decade or so, people in developed parts and urban centres around the world have become used to relatively cheap, high-bandwidth, always-on connectivity to the Internet. Consequently, many Internet applications and security solutions for them presume the availability of immediate end-to-end connectivity between communicating nodes, as well as to security infrastructure nodes like key distribution centres or certification authorities. However, there are large parts of the world where immediate end-to-end connectivity cannot be assumed for a variety of reasons: for example, in rural areas of the developing world, the infrastructure for such connectivity is simply not available and the cost–revenue ratio of deploying the needed infrastructure is likely to be too high.

A number of recent research projects have been developing and piloting 'delay-tolerant networking' (DTN) solutions for connectivity in such delay- or disruption-prone environments. Examples include DakNet [DakNet], a system for facilitating inexpensive digital communication for remote rural areas by making use of a mobile access point mounted in a bus.

Security in DTNs have several challenges. Two of the primary challenges are:

- End-to-end confidentiality: in well-connected networks end-to-end confidentiality can be achieved by using encryption. The key for encryption may be symmetric or asymmetric. In both cases, the encryption key is typically fetched on-demand from a server: a key distribution server in the case of symmetric encryption, or a directory server containing public key certificates in the case of asymmetric encryption. Keys may be cached, but even then it is prudent to connect to an online server to check revocation status.

Obviously, in DTN scenarios on-demand access to a server cannot be assumed

• Publicly verifiable authentication: successful operation of DTNs rely on the co-operation of participating nodes: a node should be willing to expend its own resources to carry and forward messages on behalf of other nodes. To protect their resources from being abused, nodes have to use some form of resource management scheme, which relies on identifying and authenticating messages. Since any intermediate node should be able to authenticate a message, authentication must necessarily be based on asymmetric key algorithms.

A recent paper [Asokan07] outlines how GAA can be used to address these problems. End-to-end confidentiality can be addressed in two alternative ways. One is using identity-based encryption. In this case, a trusted server known as the 'private key generator' (PKG) is required to periodically provision decryption keys to recipients. The other is by using the public key of a trusted server which is required to authen ticate the recipient before delivering the decryption key. The former is suitable if both the sender and the recipient are in DTN environment, while the latter is suitable if only the sender is in a DTN environment. In both cases, the delivery of decryption keys to the recipient can be secured using GAA, where the trusted server acts as a NAF. If the receiver is also in a DTN environment, it cannot use the standard inter-active bootstrapping procedure of GAA. However, future versions of GAA will include a 'push' variant which allows bootstrapping to be done noninteractively (see section 6.1.1).

Publicly verifiable authentication can be addressed by signing messages and sending them along with GAA subscriber certificates. Since subscriber certificates are short-lived, verifiers do not need to have online access to a revocation server. If the sender is in a DTN environment, there must be a way to periodically push subscriber certificates. The current GAA subscriber certificate specification requires a fresh bootstrapping as a prerequisite to certificate issuance. However, this can be relaxed since the Certificate Authority NAF can check the validity of the subscription without requiring the UE to perform GAA bootstrapping.

4.2.4 Terminal to Terminal Security

The Technical Specification [TS33.259] describes how to set up a secure channel between a GAA-enabled device and a remote device. The remote device does not necessarily need to be GAA-enabled; it can be a PC or just some other remote device with IP connectivity. The first device will use GAA to set up a secure channel with a key distribution server called the 'NAF Key Centre'. The second device is expected to use some other procedure to set up a secure channel with the same server.

Although TS33.259 does not discuss key agreement between two GAA-enabled devices, it can be easily extended to cover this scenario. Such a key agreement will have many interesting uses. For example, it can be used to set up a shared key to enable access-independent confidential messaging between two users: one user can choose the entry for the second user from his address book and select an option to set up a security association between the two users. This will trigger key agreement protocol to be carried out which will result in a security association between the two devices. Once the security association is set-up, messages exchanged between the users can be authenticated and encrypted automatically. In this section, we outline key agreement protocols that can be used between two GAA-enabled devices.

Suppose the two devices are capable of direct communication. The responding user has a device ME1 and an associated GBA module with a cellular identity ID1. The initiating user has ME2 and ID2, respectively. Figure 4.15 illustrates the protocol for establishing a shared key Ks_shared_12 between the two devices.

1. ME2 sends a request to ME1 for a key identifier (B-TID). Since the ME2 does not know which NAF Key Centres ME1 supports, it adds a list of acceptable Key Centres and its identifier ID2. ME1 selects acceptable Key Centre from the list received from ME2.
2. If ME1 does not have a valid GAA master session key, it performs a new bootstrapping procedure according to [TS33.220] using one of the variants of GBA, which results in security association with the NAF Key Centre, represented by <B-TID1, Ks_NAF1>.
3. ME1 sends a key agreement request identity to the Key Centre. The request contains ID2 and B-TID1. NAF Key Centre will choose a shared key Ks_shared_12 to be used between ID1 and ID2. This can be random or derived from Ks_NAF1 in some deterministic manner.
4. NAF Key Centre will include Ks_shared_12 in the reply, protected by Ks_NAF1.
5. ME1 sends NAF_ID and B-TID1 to ME2 using the direct connection.
6. If the ME2 does not have a GAA master session key, it performs a new bootstrapping procedure in the same manner as ME1 did before.
7. ME2 can now request the shared key by sending B-TID1 and ID2 to the NAF Key Centre-
8. NAF Key Centre will include Ks_shared_12 in the reply, protected by Ks_NAF2.

Now both ME1 and ME2 have a security association in the form of the shared key Ks_shared_12.

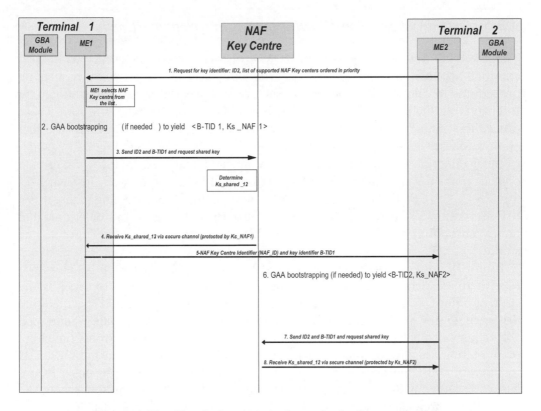

Figure 4.15. Terminal-to-terminal security in direct pull mode

This variant can be called the 'direct pull mode' variant because the initiator (terminal 2) causes the responder to begin the key agreement process with the NAF Key Centre and send the resulting key identifier (B-TID1) back to the initiator. An alternative would be for the initiator to contact the NAF Key Centre first. As a result, the shared key will be associated with the bootstrapping identifier B-TID2 of terminal 2, which can then be pushed to the responder. This can be called the 'direct push mode'.

A third variant can be the 'indirect' mode: the initiator contacts the NAF Key Centre and requests a shared key with the responder ID1. The procedure looks similar to the direct push mode, except that the responder does not get the shared key Ks_ shared_12 right away. When the responder accesses the NAF Key Centre the next time, the Key Centre can indicate that it has a waiting shared key with ID2. If the responding user decides to accept the key, it can be sent via a GAA-protected channel. This way, the two devices can receive a shared key without having any direct communication during the key agreement process.

4.2.5 Transitive Trust in IP Multimedia Subsystems (IMS)

The evolving interworking of the IP Multimedia Subsystem (IMS) with open external applications lead to new security questions for IMS. 3GPP agreed on procedures to allow a SIP application server to request access to IMS through an Interrogating Call Session Control Function (I-CSCF) on behalf of a user. These procedures include routing of the SIP request through an I-CSCF in the case that application server does not have knowledge of the selected Serving Call Session Control Function (S-CSCF), e.g., in the unregistered case. An I-CSCF serves as a contact point towards other network operators and hides the network internal IMS topology. A S-CSCF is the central network node in the IMS signalling plane, it is a SIP server, but also performs the session control (including decisions on routing and policy enforcing) and interacts with the HSS where the relevant user profiles are stored.

If an SIP application server sends an SIP request to an I-CSCF on behalf of the user, it is as such not authenticated since the I-CSCF is designed to serve as the contact point for other operator networks. To solve this problem, it is specified that the I-CSCF may maintain a list of trusted application servers that are allowed to send SIP requests on behalf of the user. This implies that the I-CSCF performs some kind of filter functionality on incoming traffic and this also implies that the presented identity of the application server needs to be verifiable. One approach is to use Network Domain Security Protocols (NDS/IP) [TS33.210] to protect the traffic between the application server and the I-CSCF. Alternatively, if the I-CSCF should not be burdened too much, the S-CSCF could be aware of which application servers are trusted. Then it verifies the requests by checking that the I-CSCF and application server headers are correct and trustworthy. These measures provide good security if the application server is under control of the network operator.

The situation gets more complex when the application server is *not* under operator control, but instead a third-party server. In this case, the measures above provide some security level, but additionally, the network operator needs to secure the pipe to the application server, e.g., using TLS [RFC2246]. To ensure well behaving of the application server, contractual agreements are recommended between the application service provider and the network operator.

When there are only few service providers, then the measures above, i.e., white lists, certificates and contractual agreements, should work reasonably well. But the numbers of standardization bodies using IMS are growing and besides the classical MNOs also include now cable network providers [CableLabs] and fixed network providers Telecoms & Internet converged Services & Protocols for Advanced Networks [TISPAN]. Hence, it can be expected that with the roll-out and deployment of IMS, the amount of application servers and the range of offered services will increase rapidly. This implies that the management of the application servers in the I-CSCF might get very inefficient and expensive. Also not all application services connected into those networks may be fully trusted. From application services' point of view, the registration with several MNOs may be cumbersome and time-consuming.

From the user's perspective, the scenario becomes worrying if there is an application service that is setting up a session on his behalf, but the user has actually not requested any kind of service. This then may lead to a situation that either the user is charged incorrectly or that the home operator of the user has to suffer the costs. The situation is similar to roaming fraud or premium service fraud in roaming scenarios, which cause MNOs quite some trouble.

Another worry of the network operators is that such a setting-up of sessions on behalf of the user from an 'assumed' trustworthy node may congest his network nodes and lead to Denial of Service (DoS) attacks. The 'assumed' trustworthy node may not be what it claims to be and uses spoofing methods to hide his own identity, like spammers do today in the Internet. How can the I-CSCF (or S-CSCF) be sure that an incoming request is really coming from one of its own nodes or partners?

As a summary, we can conclude that the current security measures are sufficient, if there are few services and the trust link between operator and service is very close, but with the opening up of the networks and potentially larger amount of service, additional security means should be considered and studied.

We will now present a sketch of how GAA could be utilized to make sure that the service requests were really made on behalf of a user. Let us start with the most obvious question:

What kind of services may have the need for setting up a SIP session on behalf of a user?

The scenario that an SIP session is set-up on behalf of a user is not so unusual as one might first think. Typical examples include:

Conferencing application:
A conferencing application that sets up conference calls (maybe even with video- and data-sharing functionality) at a prescheduled time. At the time of the conference call, the participants are then invited to join the call. This application of special interest in international companies, where international roaming with different operators is needed.

Greeting card service:
A greeting card service that sends out multimedia greeting cards at a time that is specified by the user. Here, it is quite natural that the user does not need to send out the card immediately when contacting the service. Also this scenario may involve that the greeting card service has to interact with many different operators to provide good coverage for his service.

Task manager:
Many people use task manager that sends out reminders, e.g., sending out a reminder for a group task. This could be realized by sending out a scheduled SIP message on behalf of one person. Examples of reminders in this category are for remembering to fill in the working time into a tool or that a password needs to be changed.

Betting service:
A user might want to preschedule a bet at a specified time. There exist some solutions already, but they are usually part of the betting application itself, and hence, of the corresponding application silo. For interoperability, it may be of general interest to be able to set up a predefined SIP message for a specific time.

All examples above have the property that the application server may not contact the I-CSCF directly when the user requests the service at the application server (i.e., there is a time gap between actual user authorization and SIP request message). Also, it might be that the user is not reachable at the time when the SIP request is sent, which implies that the home network can not ask the user to confirm the correctness of the request. Imagine you want to send an electronic birthday greeting card to a friend in another time zone and your operator calls you in the middle of the night to ask if you really wanted to send this card? This example illustrates why the straightforward solution just to prompt the user to confirm the request on his behalf is not user-friendly. Also, the user may not be reachable at all because of, e.g., phone battery being empty, phone switched off, no coverage, etc. From the security point of view, prompting the user with security questions is not a very successful approach since very few people really read the security warnings properly before pressing 'yes'.

Therefore, an approach that generates an authorization token that can be verified by the IMS network would be logical. For this authorization token approach, GAA could be utilized. The terminal could bootstrap AKA, and the S-CSCF acting as NAF could obtain the shared secret from the BSF. Then the terminal and the S-CSCF both generate an authorization token starting with data containing, e.g., service provider name, time, service type and encrypting the data with the GAA key. The token would then be transferred to the application service that includes it in the SIP request. The I-CSCF would either fetch then the token from the S-CSCF for verification or forward the SIP request containing the token to the S-CSCF for verification. This approach works if the SIP request is sent directly after the service request and bootstrapping run, else it might be possible that the token is reused in an unauthorized manner or a replay attack is performed. For prescheduled services, a slight variation might be interesting.

GAA could be used also to secure the tunnel between the application server and the terminal, in which case, the application server would act in GAA-terms as

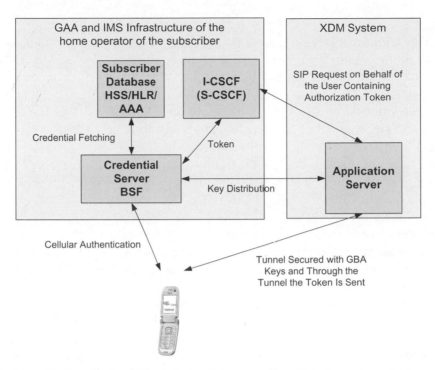

Figure 4.16. IMS authorization on behalf of the user using GAA. Reproduced by permission of © Nokia

NAF. The authorization token is sent through this secure tunnel. This alternative is depicted in Figure 4.16:

In general, introducing GAA for authorization purposes in the context of the IMS network has some impacts that need to be considered carefully.

- I-CSCF is a stateless SIP Proxy in the IMS system. If token validation functionality would be added, then this situation may need to change. It can be considered to move this token comparison to the S-CSCF, but then the I-CSCF needs to transport the token to the S-CSCF and wait for the answer. The return, e.g., granting / rejection of service request may utilize some existing mechanisms, but the transport to the S-CSCF most likely needs some additions to the existing protocols.
- Modification of SIP messages or reusage of existing fields to transport the authorization token from the application server to the I-CSCF (or even to the S-CSCF) requires most likely standardization in IETF or integration into the ongoing [IETF-SIP-SAML] work. Alternatively, there could be a separate message form the application server, but then we would need some form of binding between this request message and the token using, e.g., SIP and SAML. Or this binding could

be achieved by hashing the SIP request message and sending the hash value together with the token in a secure tunnel (TLS or IPSec) to the IMS network.

- Usage of the token: The token could be a one-time token, a time-limited token or a sort of general authorization or blanket token. The choice has to be studied carefully with regard to the use cases, but one also needs to consider potential attacks, like replay attacks or attacks by a malicious service provider. The storage of the token and keys may offer an alternative, but may also cause problems, especially if only a one-time token is actually desired.
- Convergence: GAA was designed for mobile networks and the presence of a smart card in the requesting device, but there are approaches and usages also in fixed networks, e.g., [TISPAN], [CableLabsGBA].

On the other hand, introducing an authorization mechanism to authorize service requests on behalf of the user has by nature always some sort of larger impact and the GAA approach would at least utilize an infrastructure that is already used by many other services.

Currently, some work is ongoing in IETF how a SAML assertion could be placed into an SIP request for authorization purposes [IETF-SIP-SAML]. This work has a two-phase approach: first the caller obtains an SAML assertion and then the assertion is presented to a verifier. Both phases are SIP-based and the caller and the verifier have no direct communication link in the current architecture. The authenticating entity forwards the SIP request from the caller to the verifying party. Hence, the architecture scenario is slightly different and can not be applied directly to the use cases we have outlined above, but future enhancements may benefit from it. Also, this method does not provide direct means to integrate the current cellular authentication, but that could be hidden in the SAML assertion.

References

[Asokan07]   N. Asokan, K. Kostiainen, P. Ginzboorg, J. Ott, C. Luo, *Applicability of Identity-based Cryptography for Disruption-tolerant Networking*, Proceedings of the 1st international MobiSys Workshop on Mobile Opportunistic Networking. Pp 52–56, 2007. Available at http://doi.acm.org/10.1145/1247694.1247705

[CableLabs]   CableLabs – PacketCable. Available at http://www.packetcable.com

[CableLabsGBA]   CableLabs – PacketCable, PKT-SP-33.220-I02-061013, PacketCable 2.0 IMS Delta Specifications *Generic Authentication Architecture (GAA); Generic Bootstrapping Architecture Specification* 3GPP TS 33.220, Version 2.0, October 2006. Available at http://www.packetcable.com/downloads/specs/PKT-SP-33.220-I02-061013.pdf

[CableLabsZh]   CableLabs – PacketCable, PKT-SP-29.109-I01-060914, PacketCable 2.0 IMS Delta Specifications *Generic Authentication Architecture (GAA); Zh and Zn Interfaces Based on the Diameter Protocol; Stage 3 Specification* 3GPP TS 29.109,

	Version 2.0, September 2006. Available at http://www.packetcable.com/downloads/specs/PKT-SP-29.109-I01-060914.pdf
[DakNet]	A. Pentland, R. Fletcher and A. Hasson, *DakNet: Rethinking Connectivity in Developing Nations*, IEEE Computer, 37:1 (78–83), January 2004. Available at http://doi.ieeecomputersociety.org/10.1109/MC.2004.1260729
[DVB-H]	http://www.dvb-h.org/
[IETF-SIP-SAML]	Internet Engineering Task Force (IETF), *SIP SAML Profile and Binding*, draft-ietf-sip-saml-02, May 2007. Available at http://www.ietf.org/internet-drafts/draft-ietf-sip-saml-02.txt
[Martin04]	Thomas Martin, *A Set Theoretic Approach to Broadcast Encryption*, Ph.D. Thesis, University of London, 2004.
[OBCPoster]	N. Asokan and J.E. Ekberg, *A Platform for OnBoard Credentials*, In the Proceedings of the Twelfth International Conference on Financial Cryptography and Data Security (Financial Crypto), January 2008.
[OBCTR]	Jan-Erik Ekberg, N. Asokan, K. Kostiainen, P. Eronen, A. Rantala and A. Sharma, *OnBoard Credentials Platform Design and Implementation*, Nokia Research Center Technical Report NRC-TR-2008-001, January 2008. Available at http://research.nokia.com/tr/nrc-tr-2008-001
[OMA]	Open Mobile – Alliance (OMA). Available at http://www.openmobilealliance.org/
[OMAGBAProf]	Open Mobile Alliance (OMA), OMA-Draft-TS-GBA-Profile-V1_0-20070607-D, *OMA GBA Profile*, Draft Version 1.0, 07 June 2007. Available at http://member.openmobilealliance.org/
[OMAPres]	Open Mobile Alliance (OMA), OMA-TS-Presence_SIMPLE-V1_0-20060725-A, *OMA Presence Simple Specification*, Approved Enabler Version 1.0, July 2006. Available at http://www.openmobilealliance.org/
[OMASA]	Open Mobile Alliance (OMA), OMA-AD-BCAST-V1_0-20070529-C, *Architecture of the Mobile Broadcast Service*, Candidate Version 1.0, 29 May 2007. Available at http://www.openmobilealliance.org/
[OMASC]	Open Mobile Alliance (OMA), OMA-TS-BCAST_SvcCntProtection-V1_0-20070529-C, *Service and Content Protection for Mobile Broadcast Services Specification*, Candidate Version 1.0, September 2007. Available at http://www.openmobilealliance.org/
[OMASUPL]	Open Mobile Alliance (OMA), OMA-AD-SUPL-V1_0-20070615-A, *OMA Secure User Plance Location Architecture Specification*, Approved Enabler Version 1.0, June 2007. Available at http://www.openmobilealliance.org/release_program/docs/SUPL/V1_0-20070615-A/OMA-AD-SUPL-V1_0-20070615-A.pdf
[OMAXDM]	Open Mobile Alliance (OMA), OMA-TS-XDM_Core-V2_0-20070724-C.pdf, *XML Document Management Specification*, Version 2.0, July 2007. Available at http://www.openmobilealliance.org/
[RFC0768]	Internet Engineering Task Force (IETF), *User Datagram Protocol (UDP)l*, RFC 768, August 1980. Available at http://www.ietf.org/rfc/rfc0768.txt
[RFC0793]	Internet Engineering Task Force (IETF), *Transmission Control Protocol (TCP)*, RFC 793, September 1981. Available at http://www.ietf.org/rfc/rfc0793.txt
[RFC2246]	Internet Engineering Task Force (IETF), *The TLS Protocol Version 1.0*, RFC 2246, January 1999. Available at http://www.ietf.org/rfc/rfc2246.txt

[RFC2617]   Internet Engineering Task Force (IETF), *HTTP Authentication: Basic and Digest Access Authentication*, RFC 2617, June 1999. Available at http://www.ietf.org/rfc/rfc2617.txt

[RFC3261]   Internet Engineering Task Force (IETF), *SIP: Session Initiation Protocol*, RFC 3261, June 2002. Available at http://www.ietf.org/rfc/rfc3261.txt

[RFC3546]   Internet Engineering Task Force (IETF), *TLS Extensions*, RFC 3546, June 2003. Available at http://www.ietf.org/rfc/rfc3546.txt

[RFC3550]   Internet Engineering Task Force (IETF), *Real-time Transport Protocol (RTP)*, RFC 3550, July 2003. Available at http://www.ietf.org/rfc/rfc3550.txt

[RFC3830]   Internet Engineering Task Force (IETF), *MIKEY: Multimedia Internet KEYing*, RFC 3830, August 2004. Available at http://www.ietf.org/rfc/rfc3830.txt

[RFC4279]   Internet Engineering Task Force (IETF), *Pre-Shared Key Ciphersuites for Transport Layer Security (TLS)*, RFC 4279, December 2005. Available at http://www.ietf.org/rfc/rfc4279.txt

[RFC4825]   Internet Engineering Task Force (IETF), *The Extensible Markup Language (XML) Configuration Access Protocol (XCAP)*, RFC 4825, May 2007. Available at http://www.ietf.org/rfc/rfc4825.txt

[TISPAN]   Telecoms & Internet converged Services & Protocols for Advanced Networks (TISPAN). Available at http://www.etsi.org/tispan/

[TR33.980]   3rd Generation Partnership Project (3GPP), Technical Report TR 33.980, *Liberty Alliance and 3GPP Security Interworking; Interworking of Liberty Alliance Identity Federation Framework (ID-FF), Identity Web Services Framework (ID-WSF) and Generic Authentication Architecture (GAA)*, Version 7.4.0 (2007). Available at http://www.3gpp.org/

[TS24.109]   Generation Partnership Project (3GPP), Technical Specification TS 24.109, *Bootstrapping Interface (Ub) and Network Application Function Interface (Ua); Protocol Details*, Version 7.5.0 (2006). Available at http://www.3gpp.org/

[TS29.109]   3rd Generation Partnership Project (3GPP), Technical Specification TS 29.109, *Generic Authentication Architecture (GAA); Zh and Zn Interfaces Based on the Diameter Protocol*; Stage 3, Version 7.5.0 (2006). Available at http://www.3gpp.org/

[TS33.141]   3rd Generation Partnership Project (3GPP), Technical Specification TS 33.141, *Presence Service, Security*, Version 7.1.0 (2006). Available at http://www.3gpp.org/

[TS33.210]   3rd Generation Partnership Project (3GPP), Technical Specification TS 33.210, *3G security; Network Domain Security (NDS); IP Network Layer Security*, Version 7.2.0 (2006). Available at http://www.3gpp.org/

[TS33.220]   3rd Generation Partnership Project (3GPP), Technical Specification TS 33.220, *Generic Authentication Architecture (GAA); Generic Bootstrapping Architecture*, Version 7.7.0, (2007). Available at http://www.3gpp.org/

[TS33.221]   3rd Generation Partnership Project (3GPP), Technical Specification TS 33.221, Generic Authentication Architecture (GAA); *Support for Subscriber Certificates*, Version. 6.3.0 (2006). Available at http://www.3gpp.org/

[TS33.222]   3rd Generation Partnership Project (3GPP), Technical Specification TS 33.222, *Generic Authentication Architecture (GAA); Access to Network Application Functions Using Hypertext Transfer Protocol over Transport Layer Security (HTTPS)*, Version 7.2.0 (2006). Available at http://www.3gpp.org/

[TS33.223]   3rd Generation Partnership Project (3GPP), Technical Specification TS 33.223, *Generic Authentication Architecture (GAA); Generic Bootstrapping Architecture (GBA) Push Function*, Version 0.1.0 (2007), Release 8.

[TS33.246]   3rd Generation Partnership Project (3GPP), Technical Specification TS 33.246, *3G Security; Security of Multimedia Broadcast / Multicast Services*, Version 7.3.0 (2007).

[Venyon]   http://www.venyon.com

5

Guidance for Deploying GAA

5.1 Integration with Application Servers

This section will discuss how an application server may be converted to be GAA-aware and be enabled to use GAA for authentication.

5.1.1 Introduction

In order for an application server to authenticate subscribers using GAA, and possibly, also otherwise secure the communication between the terminal and the application server it must be GAA-aware. In practice, this means that the application server must contain the NAF functionality, which enables it to:

- request the GAA parameters from the BSF, and
- process the communication between the terminal and the server to utilize these GAA parameters.

There are two distinct mechanisms to implement this. An application server developer may use a ready-made fully configurable NAF library that embeds all necessary and potentially complex communication mechanisms between the NAF and the BSF as well as provide general utility functions to handle the requests incoming from the UE. The other mechanism is to use the Web Service Description Language (WSDL) description of the web services-based Zn interface to generate all the necessary

Cellular Authentication for Mobile and Internet Services
Silke Holtmanns, Valtteri Niemi, Philip Ginzboorg, Pekka Laitinen and N. Asokan
© 2008 John Wiley & Sons, Ltd

skeleton code from the NAF, which in this case would be just a Web Service Client or Consumer. These approaches are discussed more in detail from Section 5.1.3 onwards.

It is also possible to add GAA awareness to an application server transparently. In this case, the application server is not modified at all but either the authentication server is modified to support GAA-based credentials or a proxy can be introduced between the application server and the authentication server that would handle the GAA-based authentication. This approach is discussed in more detail in Section 5.1.2.

5.1.2 Username / Password Replacement

Most of the application servers in the Internet today base their end-user authentication on username / password schemes. Typically, the user is presented a dialog in a dedicated application or an HTML form in a web browser to which the user is able to write his username and password. This credential data is then transferred to the application server by various means, e.g., using HTTP secured by server-authenticated SSL or TLS tunnel, also known as HTTPS. To enable GAA-based authentication in this scenario, the application in the client side must be made GAA-aware and it would use the B-TID as the username, and the application-specific GAA key (Ks_NAF) at the password. In practice, the client application should be capable of recognizing when GAA-based credentials are needed, be able to fetch the credentials from the GAA module and insert them to the correct context. With the web browser and with HTML forms, this would mean that, for example, javascript is enhanced with capabilities for such functionality.

There are two options for adapting an existing application server. It can either be modified to support GAA directly or the password validation procedure can be tweaked to support GAA-based credentials. The former case is discussed later in this section. The latter case is more elaborated here.

Figure 5.1 presents an example scenario where a proxy has been introduced. The system and the split-terminal scenario is used where GAA credentials would become available on the PC side over a local interface, such as over Bluetooth.

Typically, application server validates username/passwords using LDAP- (Lightweight Directory Access Protocol) or RADIUS-based interface, for example. A developer could introduce proxy functionality between the application server and the authentication server. This proxy would intercept all authentication requests coming from the application server, and in case they are GAA-based credentials, process them locally. If the credentials are normal username / passwords, the authentication request is forwarded to the authentication server. If the credentials are GAA-based, the proxy would use the B-TID in the username field of the authentication request

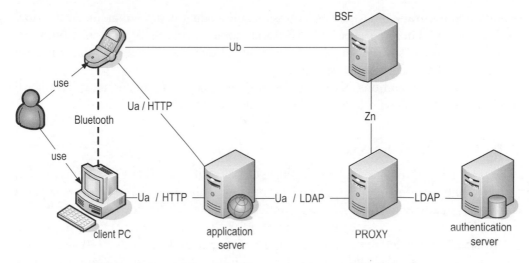

Figure 5.1.   Adding GAA support without modifying servers

and fetch the Ks_NAF and User Security Settings (USS) from the BSF. Then it would check that Ks_NAF in the password field is the same as the Ks_NAF received from the BSF. Alternatively, if the authentication request contains a response to a challenge, the proxy would validate the response using the Ks_NAF. In any case, if the validation of the authentication request is successful, the proxy will return an authentication success message back to the application server.

As the application server may need to access user-specific data in its local databases, it may be necessary to change the interface to that database in such a way that the user identity returned by BSF B-TID is added as a key to access the data: A query with B-TID as the user name would be handled by the proxy instead of the local database. This way the application server can fetch normally the user data using the B-TID that it thinks to be the username.

This could be seen as the first step towards making an application server GAA-aware. An advantage of this approach is that the normal username / password scheme would work alongside the GAA-based scheme, and thus, GAA-based usage from a mobile phone and username / password-based usage from a PC would be possible for the same end-user.

5.1.3 NAF Library

The NAF library may come in many different forms. Its aim is to ease the development of the NAF application server logic and especially minimize the code lines dedicated to the GAA-specific functionality. In the ideal situation, all changes are done in configuration files and no GAA-specific coding is needed at all. At the time of writing this book, no general NAF library tools are publicly available, but most likely BSF vendors will provide tools for application server developers that will ease utilization of the GAA infrastructure.

The following sections give some examples how NAF functionality and NAF library integration could be achieved with the existing application servers.

5.1.3.1 Apache Web Server

The most straightforward way to introduce GAA-aware Apache web server [Apache] is to enhance the commonly used Apache authentication module mod_auth to support GAA-based authentication. As mod_auth already contains support for HTTP Digest, it would be relatively easy to add the support for it to recognize and handle the HTTP Digest challenges and request based on GAA as specified in [TS24.109]. When this support would be available in the authentication module, it could be turned on by editing the corresponding configuration files, like httpd.conf or .htaccess files. The application server logic itself would then get a hold of the authenticated GAA identity through an environment variable.

Below is an example of what an *.htaccess* file could contain:

```
# GAA authentication module settings
AuthType GAADigest
AuthName "3gpp-bootstrapping@example.server.com"
BSFAddress "bsf.operator.com"
BSFConnType WebService
BSFGSIDSet 1,9999
```

The AuthType is set to be GAADigest to indicate to the authentication module that the authentication of the requests concerning this resource should be based on GAA-based HTTP Digest. The AuthName is set as specified in [TS24.109] to contain the indication to use GAA-based authentication ('3gpp-bootstrapping@' prefix) and the NAF DNS name as the domain part of the AuthName. The BSFAddress would define where the default BSF is located and the BSFConnType instructs the authentication module to communicate using web services-based Zn interface. Finally, the BSFG-SIDSet defines the GSIDs that the authentication module should add to the request outbound to the BSF. The BSF would then return those USSs that it could find from subscriber's GUSS.

The above would just enforce that GAA authentication is required for the particular resource. The application server developer can be sure that the end-user is a valid subscriber of the operator. If the application logic wants to enforce a user-specific access control, then it must retrieve the specific persistent identity of the subscriber from the authentication module. This can be done using environment variables, where there could be variable called 'GAA-Identity' which would contain the identity, such as an IMSI, provided by the BSF. The authentication module would handle the details from where the identity would come, e.g.: Is it the IMSI, an identity from a USS, or set of identities originating from single or different USSs elements? The application logic would then just use the 'GAA-Identity' to access to persistent identity information of the authenticated subscriber.

5.1.3.2 J2EE Servers

Java 2 Enterprise Edition [J2EE] could embed the GAA-based authentication to a filter that handles incoming requests. In principle, this would work the same way as in the Apache web server case, where a GAADigestAuthenticationFilter would have been implemented as part of the NAF library. The GAA filter would be configured in web.xml file of the corresponding J2EE application.

Below is an example what a *web.xml* file could contain:

```
<!- GAA authentication filter ->
<filter>
  <filter-name>GAAFilter</filter-name>
  <filter-class>com.nokia.gaa.filter.GAADigestAuthenti
cationFilter</filter-class>
  <init-param>
      <!- put here the URL to the  BSF ->
      <param-name>bsfURL</param-name>
      <param-value>http://bsf.operator.com:5320/</param-
value>
  </init-param>
</filter>
<filter-mapping>
  <!- put here the url-pattern for which GAA
authentication will be used ->
  <filter-name>GAAFilter</filter-name>
  <url-pattern>/gaa-protected/*</url-pattern>
</filter-mapping>
```

A filter, called GAAFilter in the above web.xml file, is configured to handle incoming requests for resources whose URLs match the pattern '/gaa-protected/*': All requests to access the '/gaa-protected/' directory or any of it contents would be authenticated using GAA. The parameter 'bsfURL' defines the URL to access the BSF using the web services-based Zn interface. The application logic itself can use session properties to convey the persistent identity of the authenticated subscriber received from the BSF.

5.1.3.3 Direct Usage of NAF Library

The NAF library could be used also directly because NAF libraries would offer rich set of APIs to interwork with the GAA-based parameters and functions. Below is an example of java code that utilizes NAF library API directly:

```java
// import the package
import com.nokia.research.gaa.*;

// class definitions, etc. in here.

// Get B-TID from incoming request
String btid = request.getParameter("btid");

// Initialize the connection to BSF
GAAConnection c = GAAFactory.getGAAConnection(
                  "com.nokia.gaa.internal.
SOAPConnection",
                  "http://bsf.operator.com:9999/");

// Create the request, happens typically when
// UE has contacted NAF and given B-TID
GAARequest req =
   GAAFactory.getGAARequest(btid,"example.server.com");

// Add one GSID to request certain USS
req.addAid("123");

// Send the query to BSF
GAAResponse resp = c.send(req);

// Get persistent user identity (this would be the IMPI):
String uid = resp.getUid();
// Get the ME key (normal GAA):
byte[] key = resp.getMEKey();
// use the key as it is (e.g., in HTTP Digest as
password)
```

```
// Get application specific USS (if not found this
// returns null):
USS uss = resp.getUSS("123");

// Test if it contains an authorized UID allowed by
access
// control policy (uid, which is fetched from local
policy
// system of the NAF).
if (uss.containsUid(uid)) {
  // yes it does
}

// any additional service logic follows
```

In this example, the B-TID is extracted from the incoming request and a handle to a connection to the BSF is obtained. Then the request for GAA credentials and one USS is created and sent to the BSF. After this the response is received from the BSF including the authenticated identity of the subscriber and the particular USS that was requested. The ME key (i.e., GBA_ME Ks_NAF key) is used to authenticate the incoming request in whatever way the application logic requires.

5.1.4 Web Services Direct Usage

As with any web services-based development tools, the application developer may also generate the basic NAF functionality by just using the WSDL definition file for the Zn interface. This definition file is defined in [TS29.109], Annex D. It can be used in any web services-based tools to generate the client-side skeleton code to access the BSF. In this case, the developer must also implement all other supporting functionality as well because WSDL describes only the interface itself. This is a suitable approach in cases where no ready-made NAF library is available: for example, the application server may be implemented in programming language, which supports web services but does not yet have an NAF library implemented in that particular programming language.

5.2 Integration with OS Security

In this section, we will examine the security issues involved in the UE-side implementation of GAA and how they are addressed.

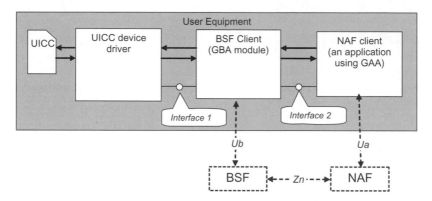

Figure 5.2.   UE-side software architecture for GAA

5.2.1 Threats for GAA Implementations in Open Platform UEs

Figure 5.2 shows a typical software architecture for a UE-side GAA implementation. There are two interfaces in the architecture which may call for access control. One is the interface to the UICC (via a device driver). The other is the interface to the BSF client, which is also called the 'GBA module'. In closed platforms, all the software modules on the UE are preinstalled and can be deemed trusted. Therefore, access control is not an issue. On the other hand, the inherent feature of open platforms is that new applications can be installed on the UE. On such open platforms, the following threats arise:

- **Interface 1:** A malicious application can access the UICC directly, and therefore, can function as a BSF client, which can then communicate with both the UICC and the BSF, and hence, establish the GAA master session key.
- **Interface 2:** The malicious application can access the BSF Client API and obtain NAF-specific GAA session keys from it.

The malicious application can send the stolen secrets (either the GAA master session key or one or more NAF-specific GAA session keys) to an accomplice in the network. Thus, we can imagine two types of attack: an attacker can masquerade as a NAF towards the UE or it can masquerade as a UE towards a NAF. In both cases, the attack can result in loss of private data or unauthorized usage of the service.

   We assume that the UE platform provides memory protection, isolating run-time memories of processes from one another. If the platform provides private storage for applications (like in Symbian OS), the BSF client or a NAF client can persistently store their respective secrets. Otherwise, the secrets can only be kept in run-time memory.

5.2.2 Access Control Requirements

Each of the two threats can be addressed by specifying and enforcing access control at the respective interfaces. The first threat can be addressed by restricting access to the UICC only to the authorized BSF client.

The second threat can be addressed by restricting the access to the GAA server API to authorized applications. We consider two issues with this latter access control: the setting of access control policy and the granularity of access control.

Granularity: The simplest approach is to have only two categories of applications. Authorized applications are allowed to use the BSF client API. Unauthorized applications are not. An authorized application can get the NAF-specific session key for any NAF. This is **coarse-grained** access control. Alternatively, we may want to restrict specific applications so that they can only get the NAF-specific session keys for a subset of NAFs or application protocols. This is **fine-grained** access control.

Policy-Setting: In either type of access control, the policy may be set by the device manufacturer, the network operator, a third party (such as the system administrators in an enterprise) or the users themselves.

The coarse-grained access control method is enough for applications that have been authorized by the UE manufacturer or the operator. However, if a UE allows users or third parties to authorize applications, there may be a need for a fine-grained access control since users may grant the authorization to an application simply because an application just requests it and third-party authorizers may not be trusted by one another or by the manufacturer/operator. Finer-grained access control is therefore needed to constrain the access rights of applications. For example, an operator may specify a set of sensitive NAFs such that only operator-authorized applications can get the NAF-specific keys for those NAFs. User-authorized applications could then get the NAF-specific keys for any NAF provided they are not classified as sensitive by the operator policy.

We identify two levels of access control implementations in practice. At the basic level, only coarse-grained access control is supported. Users are not allowed to authorize applications or set policy. For many scenarios and applications, this basic access control is sufficient. At the extended level, users are allowed to authorize applications and finer-grained access control is supported. Access control policies may be specified as inclusion lists (e.g., 'Application ABC is allowed to request NAF-specific keys for NAF whose identifier matches the pattern x.y.z') or exclusion lists (e.g., 'Only application ABC is allowed to request NAF-specific keys for x.y.z').

5.2.3 Basic Access Control in Practice:
Integration in the Series 60 Platform

Version 3.0 (and newer versions) of the Series 60 platform is based on Symbian OS 9.0, which incorporates platform security. We will take a brief look at the essential features of Symbian OS Platform Security before seeing how GAA implementation on the Series 60 platform makes use of it. For a comprehensive presentation of Symbian OS Platform Security, see [Heath06].

In Symbian Series 60 3.0 (and above), *all* applications that are to be installed must be digitally signed – unsigned applications cannot be installed. There are two primary groups of applications:

- **trusted applications**, which have been signed by *trusted signers* whose certificate(s) can be validated to a trusted root certificate in the device; and
- **untrusted applications**, which have been signed by *untrusted signers* whose certificates cannot be validated.

Run-time access control in Symbian platform security is based on *capabilities*. A capability is a representation of a permission to access a resource. Each process has a list of capabilities associated with it. In Symbian OS, resources are guarded by server processes. When a process requests access to a resource from a server process, the server process checks to see if the requesting process has the capabilities needed for accessing the resource.

Capabilities are granted to applications at installation time. Each software package lists the capabilities it needs for correct execution. Installation succeeds only if the Application Installer decides to grant all the requested capabilities. If the application is trusted and the requested capabilities constitute a subset of the meta-capabilities listed in the signer's certificate and in certificates that are in the certification path from signer's certificate to the trusted root certificate, then the Application Installer will grant the requested capabilities. If the application is untrusted and the requested capabilities belong to the subset of user-grantable capabilities, then the Installer may ask the user whether to grant the requested capabilities.

Figure 5.3 shows the software architecture of GAA in the Series 60 platform. The BSF client is a Symbian OS server known as the 'GAA Server' or 'GBA Module'. NAF clients can access the GAA server by using the GAA client library.

Access Control on interface 1 in Figure 5.3 is policed by the UICC 'device driver' (we use the term 'device driver' in a loose sense: depending on how the Series 60 platform is integrated into the hardware architecture, communication between the GAA Server and the UICC may traverse through several software modules). To eventually access the UICC, the GAA Server needs the ReadDeviceData and WriteDeviceData capabilities. The GAA Server also needs the NetworkServices capability in order to be able to communicate with the BSF over a network interface.

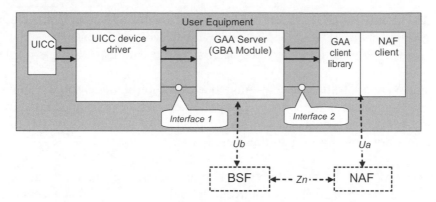

Figure 5.3. UE-side GAA software architecture in the Series 60 platform

Access control on interface 2 in Figure 5.3 is policed by the GAA Server. To be able to access NAF-specific secrets, the NAF client must have at least the Read-DeviceData capability. Additionally, the GAA Server may require a higher-level capability than the ReadDeviceData to protect more crucial NAF-specific keys. This is the case for 3GPP MBMS and OMA BCAST Smartcard profile keys, where the NAF-specific keys are used to protect the key management procedures of those systems. In these cases, the GAA server will examine the Ua security protocol identifier of the NAF_ID and if it corresponds to either 3GPP MBMS or OMA BCAST Smartcard profile identifier, the GAA server will require the NetworkControl capability, which can be issued typically only by the device manufacturer.

The NAF client may also specify the address of the BSF to bootstrap from, in which case, it needs the WriteDeviceData capability as well. NAF clients may need other capabilities, depending on what they do. For example, most NAF clients would need NetworkAccess capability to open network connections to NAFs.

5.2.4 Extended Access Control: Design Options

The previous section explained how basic access control is done in the current GAA implementation on the Series 60 platform. Now we consider how extended access control could be implemented.

Secure Identifier (SID) and Vendor Identifier (VID): Each Symbian OS application is identified by a Secure Identifier (SID). SID is a locally unique identifier. A SID can be authenticated provided that the SID is in protected range (under 0x7FFFFFFF). If an application has a protected range SID, the application must be signed by a trusted signer. Thus, the protected range SID can be used to authenticate the application itself.

Vendor Identifier (VID) is used to identify the vendor of the application. As with the protected-range SIDs, an application with nonzero VID must be signed by a trusted signer. Applications that have been signed by untrusted signers must use a VID of 0.

SID and VID can be used in fine-grained access control to identify the application and map the application to a policy. The policy can state that an application with an SID and/or VID can have access to one or more NAF-specific keys, and the other applications are denied access to these credentials.

Access control lists: In order for the UE manufacturer or operators to be able to configure the fine-grained access control in the UE in such way that access to sensitive NAF identifiers is restricted to certain applications, the GAA server must contain an Access Control List (ACL). The resources (objects) protected by the ACL are NAF-specific keys and can be identified using regular expressions on the NAF identifier. The subjects in the ACL are application identifiers, which are necessarily platform-specific. In Symbian OS, these can be SIDs or VIDs as discussed above. Device management (DM) systems, such as OMA DM, can be used to provision and update the GAA Server ACLs.

User queries for authorization: Figure 5.4 shows an example user query dialogs (a) for granting access to GAA session keys, and (b) for offering to remember the decision made by the user. The user query resembles the personal firewall functionality where the user is prompted whether a certain application can open or accept a network connection. On a UE based on Symbian OS, user queries for authorization may be used to constrain applications that already have the ReadUserData capability. Alternatively, in other platforms, user queries may be used to allow otherwise untrusted applications to access selected NAF-specific session keys.

The user prompt should show user the application name that is requesting access the GAA Server and the NAF address (in the form of a fully qualified domain name) for which it requests GAA session keys. The user is then asked whether to grant or reject the access. Figure 5.4a shows an example of user prompt. Additionally, the user could be asked whether this decision should be remembered in the future, i.e., next time the same application requests GAA session keys for the same NAF, the access is automatically granted. See an example of user query dialog in Figure 5.4b. This could be implemented as a user-specific ACL in the GAA server.

It is important to note that designing easy-to-use, and yet secure user dialogs is very challenging. Badly designed user prompts could lead the user to authorize the wrong application without intending to do so.

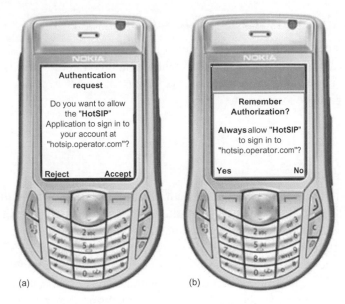

Figure 5.4. Example user query dialogs. Reproduced by permission of © Nokia

5.2.5 Other Platforms

In Section 5.2.3, we saw how access control to GAA functionality in UE is implemented using Symbian OS platform security. Similar access control can be implemented in other platforms. To implement access control, platform needs to support authentication and authorization of software modules. Typically, this has two aspects:

- Identification and authentication of software modules: there must be a way for the software platform to identify and authenticate software module, either at the time of installation or at the time of loading the module for execution (or both). Usually, this is achieved by software signing. Support for software signing and verification is included in popular operating platforms, although the extent of actual use varies greatly.
- Granting and validation of privileges to software modules: there must be a way to assign privileges to executing software modules, as well as to check privileges at the time when a software module makes a request to access a resource (like an NAF-specific key). Granting of privileges is usually done on the basis of authentication of the software module. Symbian OS supports granting and validation of privileges at the granularity of individual processes. Security Enhanced Linux (SELinux) supports similar functionality. Java security architecture provides support at a finer granularity, at the level of objects.

5.3 Integration with Identity Management Systems

The management of the personal identities has become a challenge for every consumer using a large range of services. Identity management solutions are intended to address this challenge. In this section, we discuss how GAA can be used with any identity management solution.

5.3.1 Introduction

Typically, an identity management system allows an end-user to authenticate to a central server known as the Identity Provider (IdP), and then distribute authentication tokens or assertions to other servers that want to authenticate at user. These servers are typically called either Service Provider (SP) or Relying Party (RP). The identity management specifications, in general, do not mandate a single way to authenticate to the IdP, but offer multiple authentication mechanisms as well as a way to add new authentication mechanisms. The reason for this is that it should be possible to plug a existing authentication infrastructure into identity management system. Thus, naturally, GAA can be used to authenticate a user to the IdP of a system that can be based on for example:

- Liberty Alliance Project Identity Federation Framework (ID-FF) Version 1.2 [LibertyIDFF] or Organization for the Advancement of Structured Information Standards (OASIS) Security Assertion Markup Language (SAML) Version 2.0 [OASIS-SAML], where the client application is usually a browser and does not have to be aware of the identity management system.
- Liberty Alliance Project Identity Web Service Framework (ID-WSF) Version 2.0 [LibertyIDWSF], where the client has be aware of the identity management system.
- OASIS WS-SX (Secure Exchange) [OASIS-WSSX], where the client also has to be aware of the identity management system. An example of this system is Microsoft's CardSpace [MSCardSpace].
- OpenID 2.0 [OpenID], where the client is a browser and does not have to be aware of the identity management system.

The client application has to be able to determine that GAA should be used for authentication, fetch the GAA credentials from the GBA module and use them in the particular application protocol between the client application and the IdP. For cxample, a browser can be GAA-enabled by adding the necessary hooks to the HTTP Digest authentication implementation so that it can use GAA credentials as username and password automatically when they are requested by the IdP.

In the identity metasystems like Microsoft's CardSpace [MSCardSpace] and open source projects Higgins [Higgins] and Bandit [Bandit], the card selector application needs to be aware of GAA either directly, where GAA is implemented on the same platform, or remotely, where GAA functionality is used on a mobile phone and the GAA credentials are transferred to the device over a local connection like Bluetooth or USB in the split terminal fashion as described in Section 3.5.4.

5.3.2 GAA Interworking with Liberty ID-FF

As an example of integrating GAA with an identity management system, we describe how the interworking between GAA and Liberty ID-FF works. This example is based on 3GPP TR33.980 [TR33.980], where the interworking between 3GPP GAA and Liberty ID-FF, and ID-WSF as well as OASIS SAML 2.0 is described.

In this example, the browser in the UE is GAA-aware, i.e., it is able to handle HTTP Digest authentication with GAA credentials as specified [TS24.109]. The communication between the browser and the IdP as well as between the browser and the SP is typically protected by TLS. In the sequence flow described here and shown in Figure 5-5, the TLS tunnel set-up would happen always before the first message.

1. The service user types in a URL that identifies a service that the end-user wishes to use. This service is hosted by Liberty SP. A normal HTTP GET request is sent to that URL.

    The SP examines the incoming HTTP GET request, then checks the policy of the particular service that is being accessed. The SP determines that the corresponding service has an access policy that requires users to authenticate.

    If the SP discovers that the end-user has not been authenticated yet in this session, thus decides to redirect the browser to Liberty IdP. If there is no session yet, it is created and the session information can be conveyed between the browser and the SP using an HTTP Cookie or it can be embedded to the URL used to access the services in the SP.

    If the SP discovers that the end-user has already been authenticated, steps 2–15 are skipped. This is typically done by using session identifiers embedded either in an HTTP Cookie or into the URL used for access.
2. The SP uses HTTP Response code '302 Found' directive to redirect the browser to the IdP. Besides the URL of the Liberty IdP, the Location Response Header Field also contains an embedded URL that directs the browser back to the SP service after the Liberty authentication has been completed.
3. As directed in step 2, the browser makes an HTTP GET request to the IdP.

    The IdP determines that the end-user has not been authenticated before in this session. The IdP has also been configured to use GAA for end-user

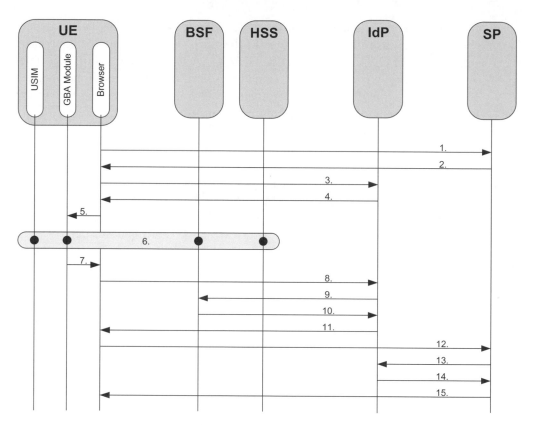

Figure 5.5.   GAA and Liberty Alliance Project ID-FF interoperation

authentication. Thus, it sends an authentication challenge in the form of HTTP Response code '401 Unauthorized' with special parameters in the WWW-Authenticate header (as specified in [TS33.222]).

If the IdP determines that the end-user has been authenticated towards IdP in this session, steps 4–10 are skipped, and instead of the '401' response as given above, the '302' response as described in step 11 is sent to the UE.

4. The browser examines the HTTP Digest authentication challenge received from the IdP and notices the special parameters in the WWW-Authenticate header. As the browser is GAA-aware, it will attempt to fetch GAA credentials (i.e., the NAF-specific key for the IdP which is acting as an NAF) from the local GBA module provided by the platform.

5. The browser makes a request to local GBA module provided by that platform to acquire the IdP-specific key needed to calculate the authentication response

for the challenge received in step 4. With this request, the browser gives as an input parameter the NAF_ID of the IdP (i.e., DNS name of the IdP and the Ua security protocol identifier) as it is used to calculate the GAA key specific to the IdP.

6. During step 6, the GBA module, together with the USIM, the BSF and the HSS, executes the GAA bootstrapping procedure if there is no valid GAA bootstrapping session.

7. In step 5, the browser requested an IdP-specific key from the GBA module. The GBA module derives it using the GAA master session key, the NAF_ID and some other parameters (see [TS33.220]). After the IdP-specific key is derived, it, along with the received B-TID and key lifetime, is given to the browser.

8. The browser calculates an authentication response to the HTTP Digest challenge received in step 4 from the IdP. It uses the B-TID as the username and the IdP-specific key (Ks_NAF) as the password in HTTP Digest calculations [TS24.109]. After the authentication response has been calculated, the browser sends an HTTP GET with the authentication response to the IdP.

9. Upon receiving the authentication response, the IdP extracts the B-TID from it, and uses the B-TID and its NAF_ID to request the IdP-specific GAA credentials (Ks_NAF) from the BSF.

   In addition to the B-TID, the IdP may include additional GAA Service Identifiers (GSIDs) to the request. A GSID identifies one particular USS in the end-user's GUSS. The USS can be used to convey, for example, an IdP-specific persistent identity to the IdP from the BSF in the case where the BSF does not wish to disclose the IMPI to the IdP.

   When the BSF receives the B-TID and the NAF_ID (and possibility one or more GSIDs), it will first locate the GAA bootstrapping session using the B-TID. If it locates the session, it will use that session's GAA master session key (Ks), the incoming NAF_ID, and some other parameters to derive the GAA NAF key (Ks_NAF) as was done by the GBA module in step 6.

   In addition to key derivation, the BSF locates the corresponding USS entries in end-user's GUSS identified by GSIDs if they were present in the request sent by the IdP. Depending on local policies in the BSF, the discovered USS entries are included in the response to the IdP.

10. The response from the BSF to the IdP includes the GAA credential, IP Multimedia Private Identity (IMPI), if the BSF policy allows it to be transferred to the Id, and requested USSs (if the BSF policy allows them to be transferred to the IdP). For details on local policy enforcement, see Section 3.5.1.

11. The IdP uses the GAA credential and the B-TID to verify the authentication response sent by the browser in step 8. If the verification is successful, the IdP constructs and digitally signs an AuthnResponse assertion that is stored locally

in the IdP. Also, an Artifact element is created, mapped to the AuthnResponse assertion and used in the next step. (The Artifact element will be used by the SP in step 13 to fetch the AuthnResponse assertion from the IdP directly.)

The IdP uses HTTP Response code '302 Found' to redirect the browser back to the SP (the URL that was embedded in messages in steps 2 and 3). This directive also includes the Artifact element itself.

12. As directed in step 11, the browser makes an HTTP GET request to the SP.
13. Upon receiving the HTTP GET request from the browser, the SP extracts the Artifact element.
14. The SP retrieves the AuthnResponse assertion from the IdP using the Artifact element. The SP verifies the AuthnResponse assertion. If the verification is successful, the SP has successfully authenticated the end-user.

The AuthnResponse assertion typically contains a pseudonym of the user that can be used in the SP as the persistent user identity. The initial request that is now authenticated and authorized is forwarded to the actual service in the SP.

15. The service returns the response to the browser determined by the executed service logic.

All subsequent requests are identified by the session information embedded into the requests. The SP will check that the session information in the requests is valid. The session is invalidated if the user has logged out from the service (at the IdP or at the SP) or the session otherwise has timed out.

5.4 Integration of GAA into Mobile Networks

In this section, we look at the issues facing a MNO who wants to deploy GAA in its networks.

5.4.1 Integration of HLR into GAA

The mobile subscriber database is one of the most important assets of an operator: its availability is crucial to provide any kind of mobile service, including voice and SMS. Therefore, it is understandable that operators are quite conservative when updating their subscriber database from an HLR to an HSS.

A full-fledged GAA including GBA_U would require an HSS with some GAA-specific functionality and support for the diameter-based Zh interface. The Zh interface is able to carry the GUSS, which is the set of all USS. The GUSS contains a field that indicates whether the subscriber has a GBA_U-enabled UICC.

This information enables the BSF to change the received authentication vector and to derive the correct keys. (See section 3.2.2) However, to generate NAF-specific keys for GBA_ME only, a BSF needs to obtain only the authentication vector from the subscriber database. This makes it possible to use an HLR or an HSS without Zh because the vectors can be requested and received over the MAP protocol. The MAP-based interface to HLR / HSS is known as Zh'. Since the MAP interface does not support GUSS and no USS storage in HLR is intended by the specification, any GAA functionality that relies on GUSS, such as GBA_U support, would not be available to a BSF using Zh'.

3GPP standardizes network nodes and protocols in Releases. GAA was specified in the Releases 6 and 7, and work has also started for Release 8 during 2007. Usually, the specifications and reports only build upon features and nodes in the same release. During creation of the specification, interoperability between releases is considered and system failures are avoided or at least caught to prevent a system breakdown. Still, real deployment and roll-out plans often look different and sometimes this is reflected in the specification work. For instance, a network operator may want to roll out Broadcast Mobile TV without updating the subscriber database to an HSS for this specific purpose or handing out new smart cards to subscribers. An update of the subscriber database would also bring 'follow-up' updates in related nodes. This poses a risk for a new service, whose take-off and success time cannot be reliably predicted. Therefore, 3GPP standardized the interworking of a Release 5 or earlier HLR, respectively of an HSS without Diameter support, with a Release 7 or Release 8 BSF in [TS33.220] and [TS29.109].

In detail, the interaction with the subscriber database is as follows: The BSF requires the authentication vector from the subscriber database to be able to derive the application-specific credentials. For this, two possibilities exist for the BSF. The mandatory diameter-based Zh interface to the HSS and the optional MAP-based Zh' interface to the HLR or to an HSS without Zh support.

The message used for the authentication vector fetching in the Zh' interface is already used for interoperator roaming or sometimes for a node internal interface, if the HSS consists of an HLR with an HSS add-on. These kinds of internal interfaces are usually not standardized. The BSF can obtain the authentication vector from the HLR by sending a MAP_SEND_AUTHENTICATION_INFO Request. The HLR would need to serve this request coming from the BSF. The HLR supports this MAP message already for the VLR node for roaming users and for the SGSN node. This MAP message is also sometimes used in an HSS–HLR node internal interface and is supported by HSS that has no Zh or Diameter support.

The HSS was basically designed for the IMS and does not necessarily support the Zh reference point or Diameter. There are some interoperability aspects between the different releases of 3GPP that are important to consider when upgrading the network to support normal GAA. These are summarized in Table 5.1:

Table 5.1. Subscriber database migration for GAA

From\To	R4 HLR with GAA	R5 HLR with GAA	R5 HSS with GAA	R6 HSS with GAA	R7 HSS with GAA
R4 HLR without GAA	Add Zh′	Upgrade to R5 and add Zh′	Upgrade to R5 HSS and add Zh′	Upgrade to R6 HSS and add Zh′ or Zh	Upgrade to R7 HSS and add Zh′ or Zh
R5 HLR without GAA	X	Add Zh′	Upgrade to R5 HSS and add Zh′	Upgrade to R6 HSS and add Zh′ or Zh	Upgrade to R7 HSS and add Zh′ or Zh
R5 HSS without GAA	X	X	Add Zh′	Upgrade to R6 HSS and add Zh′ or Zh	Upgrade to R7 HSS and add Zh′ or Zh
R6 HSS without GAA	X	X	X	Add Zh′ or Zh	Upgrade to R7 HSS and add Zh′ or Zh
R6 HSS with GAA	X	X	X	X	Upgrade to R7 HSS
R7 HSS without GAA	X	X	X	X	Add Zh′ or Zh

When the interface to HSS / HLR is based on Zh', the basic GAA functionality consisting of GBA_ME can be supported; as noted before, GBA_U cannot be supported directly and GUSS cannot be used.

5.4.2 Key Lifetime Setting in BSF

The GAA master session key Ks has a limited lifetime specified by the BSF in subscriber's home network. The ratio between key lifetime, L, and the average time between GAA service requests from a customer, T, affects how many of the GAA service requests may cause bootstrapping operations. Thus, if the home network operator is going to use L to control the average load on the BSF and HSS, he should have an idea on the range of T.

On the one hand, if L is large compared to T, then the average time between bootstrappings will be close to L. On the other hand, if L is small compared to T, then almost every GAA service request will cause bootstrapping.[1] This will not only guarantee the freshness of bootstrapped key,[2] but also increase the load on the BSF and HSS. We will now estimate the dependency of bootstrapping events on the ratio L / T under the following assumptions about service requests.

(i)   Service requests from a single customer form a sequence of independent and identically distributed random events.
(ii) Service requests by different customers are independent from each other.

Suppose that a terminal has just bootstrapped. The epoch before the next bootstrapping time starts with L seconds during which there is no need to bootstrap, followed by period during which a new service request will trigger bootstrapping. Since the length of that period is T seconds on the average, the probability p_B of bootstrapping is $T / (L + T)$,

$$p_B = \frac{1}{1 + L/T}. \tag{5.1}$$

The contribution to the HSS load by a single subscriber is p_B / T or $(L + T)^{-1}$ transactions per second. The contribution to the HSS load by N subscribers is $N (L + T)^{-1}$ transactions per second. For instance, if L is 6 hours, then 10^5 subscribers, that on the average generate four GAA service requests per hour per subscriber, will add 10^4 transactions per hour to the average load on HSS.

Figure 5.6 displays the graph of Equation (5.1): the solid line plots p_B as a function of L / T. The dots show a result of a simulation where, for each value of L / T, we

---

[1] If L is zero then every service request will cause bootstrapping.
[2] Ks will stay valid for at most L seconds after user's cellular subscription has expired.

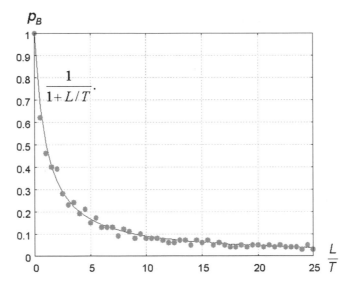

Figure 5.6. Probability of bootstrapping as a function of L / T

generated 50 exponentially distributed service requests and measured the frequency of bootstrapping events that they would have caused.

It is easy to show that without assumptions (i) and (ii) about randomness of service requests, our calculation of p_B is not valid. To illustrate, suppose first that the service requests are not independent but follow each other exactly T seconds apart. In this case, assumption (i) does not hold. If $nT < L < (n + 1)T$, where n is a non-negative number, then bootstrapping will occur every $(n + 1)T$ seconds and

$$p_B = 1/(1 + n) \tag{5.2}$$

For instance, if $n = 0$, i.e., $0 < L < T$, then $p_B = 1$, rather than $(1 + L / T)^{-1}$ predicted by Equation (5.1), because the bootstrapped key will always expire before the next service request arrives.

Second, if the time at which service is accessed matters, then many users may request the service almost simultaneously and assumption (ii) does not hold. For instance, many people are likely to start watching a real-time broadcast of sports competition at the beginning of the broadcast. A burst of service requests from many customers who do not have a valid bootstrapped key will sharply increase the load on BSF databases and Zh interface.

In summary, we recommend to use Equation (5.1) as a rough estimate of the average case and to equip the BSF and the network elements connected to it with overload control mechanisms and enough spare capacity to accommodate short temporary increases in offered load.

5.4.3 Usage of SIM Cards in GAA (2G GBA)

At the time of writing this book most subscribers have SIM cards and the roll-out of UICC cards that contain a USIM application has, for many operators, just started. Even if an operator hands out UICC cards to his subscribers, it is unlikely that he will contact his existing subscribers and ask them to replace their SIM card with a new UICC card. The costs of such a replacement are high: not only the investments in the new cards and the notifications for the replacement, but also the calls to the helpline are to be considered. Some subscribers will have problems with the physical removal and insertion of the smart card and with the potentially new PIN. This was the motivation for 3GPP to standardize the usage of the SIM card in GAA (see for technical details Chapter 3). We will now discuss the potential security risks and impacts of such a usage has and how they are mitigated.

The security level of 2G GBA, also called SIM card GBA, is different from the one of 3G GBA. The threats against 2G GBA were considered and additional protection measures were taken to prevent those attacks and to strengthen the whole overall security. From the terminal perspective, all variants of GAA are protected by the platform security. (In case of GBA_U, one key also remains in the smart card, whose access is also controlled by the operating system and the terminal platform.)

The 2G GBA solution aims to provide mutual authentication between the terminal and the BSF in the home network. We now give a brief overview of how the 2G GBA counteracts security threats related to the known GSM vulnerabilities. The threats originate from the weakness in the COMP128 algorithm and are independent of the usage of 2G GBA.

UE Impersonation

In GSM networks, the network is not authenticated; hence, there exists the threat of the impersonation of the UE to the BSF during the run of the Ub protocol. Under this category fall the following three attacks:

(1) Man-in-the-middle attack:

An attacker who owns a terminal with 2G GBA capability could try to perform a man-in-the-middle attack. In this scenario, he would impersonate a genuine 2G GBA GSM subscriber to the BSF server. The attacker would pretend to be a UE belonging to the genuine subscriber and initiate GAA bootstrapping with the BSF of the network operator. As described in Annex L of [TS33.220] the attacker would then send a challenge RAND received from the BSF to the terminal using, e.g., a false GSM base station. This way he could obtain the SRES directly, and possibly, Kc by means of cryptanalysis (if the algorithm used is vulnerable to cryptanalysis). If the victim runs the insecure A5/2 GSM encryption algorithm, then this might be feasible. It should be noted that it was intended to remove the A5/2 algorithm from the networks by the end of

2006, but since some operators have A5/2 in hardware of their base stations, the removal process may progress somewhat slowly. This attack is regarded as infeasible if one of the other GSM encryption algorithms is used.

To be able to perform this man-in-the-middle attack, the attacker would need to know that the victim user has a SIM card, is subscribed to 2G GBA and uses an old terminal with A5/2-enabled, but that still supports 2G GBA, which is a Release 7 feature. Therefore, the practical overall risk is not considered high.

(2) SIM cloning attack:

One risk that originated from the weakness of the usage of the COMP128 algorithm is the SIM cloning. Here, the attacker tries to find the cryptographic GSM key K of a genuine subscriber of a GSM network. With this key, the attacker is able to impersonate the victim completely, which then in turn includes any 2G GBA-enabled service, if the subscriber has registered to 2G GBA. If the attacker is in possession of the SIM of the subscriber he wants to attack, he potentially could also obtain the cryptographic key K by exploiting the weakness in the COMP128 algorithm that is used in insecure variants of A3/A8. This threat is not 2G GBA-specific, and operators are aware of the risk posed by using A3/A8 variants that utilize the COMP128 algorithm. Therefore, it is recommended to move towards more secure variants of A3/A8 as outlined in [TS55.205]. Then this attack will no longer be possible.

(3) Attack on UICC-ME interface:

In the mobile 2G GBA-enabled terminal, the access to the SIM card and the cryptographic details stored on it are protected by the terminal platform security. Here, the risk of compromising the smart card – terminal interface with regard to GAA can be considered equal for all 3GPP-specific variants of GAA.

BSF Impersonation

As we already mentioned earlier, in GSM networks, the network is not authenticated. Hence, we will now discuss the risk of impersonation of the BSF to the terminal during the run of the Ub protocol. In the 2G GBA solution, the BSF is authenticated to the terminal by using TLS. The attacker would need to successfully break the certificate-based TLS authentication of the BSF to the UE and mutual authentication provided by HTTP Digest using a password derived from GSM procedures. Hence, 2G GBA adds an additional layer of security here to the GSM protocols. An attacker needs to know all the parameters of the GSM triplet, in particular Kc, and additionally break the TLS security to obtain the Ks during or after the Ub protocol run. If an attacker is able to break the GSM security after the Ub protocol run, then this alone does not provide sufficient information to break 2G GBA and perform service fraud. The attacker would still need to break the TLS protocol.

Downplay Attack

Another risk that is mitigated is the so-called downplay attack, also known as bidding-down attack. In other words, a GAA-enabled terminal contains a UICC smart card with a USIM / ISIM application and a SIM application. Then the terminal could use 3G GBA based on the USIM / ISIM application on the UICC or use 2G GBA for the SIM card application on the UICC. The 2G GBA specification requires that a GAA-enabled terminal that supports SIM-based 2G GBA must also support USIM / ISIM-based GBA_ME and GBA_U for USIM. Additionally, if a USIM / ISIM is available, the terminal must use the USIM / ISIM-based GBA_ME. With this mechanism, it is ensured that if a subscriber inserts a UICC card with USIM or ISIM application, the SIM card application will not be used for service key derivation.

5.4.4 Charging and GAA

During the early phase of standardizing GAA, it was decided that authentication and authorization will be part of GAA but charging would not be. In general, in order to be able to do charging, 'triggers' would be needed: that is, certain events which could serve as a basis for the charging would trigger the generation of a charging data records. The actual monitoring and the creation of the charging record are implementation- and network-specific.

Below is a list of some example events that could be monitored:

• If one wants to charge, for example, based on the number of security associations established for a certain terminal, then the monitoring of the Ub interface between the terminal and the BSF on the BSF side is recommended.
• If the service does not reside in the home operator network, then the operator may charge the service provider, e.g., for the delivery of the GAA keys or of the USS. These events could be monitored best on BSF side to avoid that the NAF tampers with the data.
• If the charging is on the basis of service consumption, then it is part of the NAF software. The following are examples of events that can be monitored by NAFs: how many authentications of the UE to the NAF take place based on GAA keys, how many enrolments of subscriber certificates take place, how many operator root certificates were delivered secured with GAA keys, etc.
• If there is only one entity that may request GAA keys (e.g., a close roaming partner), then the Zh interface and the delivery of the authentication vector and GUSS could be monitored to be able to charge the partner. But it should be noted that this model is hard to extend, when another service provider is added.

5.4.5 GAA Integration into Large Networks

The integration of GAA into existing networks also needs consideration, particularly with regard to the special requirements of large networks or stepwise migration phases of networks.

A large network may be under the same ownership and administrative control, but still consist of different Public Land Mobile Networks (PLMNs). This may come about for a variety of reasons, such as mergers or acquisitions between operators. Optimal sharing of network resources may require that subscribers of one PLMN make use of network functions and elements in another PLMN under the same administrative control.

In GAA, the BSF is assumed to be located in the home network. Now we have the case of networks which are under one administrative control, but are split into different local networks. We now discuss two potential problems in such large networks. Consider the following scenario:

- NAF may belong to a third party network;
- Operator A hosts a BSF;
- Operator B is the home operator of a subscriber; and
- Operator A and B are under the same administrative control.

The subscriber now contacts an application server (NAF) and is triggered to make a bootstrapping run to obtain the application-specific credentials.

The Addressing and Numbering Specification [TS23.003] specifies how the terminal can derive the address of the BSF from the identity information related to the UICC application. The terminal would then use the following way to obtain the BSF address:

- If the IMPI in use is subcriber@operatorB.com, then the derived BSF address is bsf.operatorB.com.
- If the IMSI in use is, for example, 234150999999999, where the MCC is 234, and the MNC is 15, and the MSIN is 0999999999, then the BSF address is derived to be bsf.mnc015.mcc234.pub.3gppnetwork.org.

Networks without BSFs: The first problem is that not all PLMNs in the large network may have a BSF. Suppose Operator B does not have a BSF, the terminal would contact the network with one of those addresses. Now, the administrative control of the operator group could configure their DNS so that the address used by the terminal is resolved to the IP address of the BSF in the network of operator A. The terminal is not impacted by this solution.

As shown in Figure 5.7, the BSF located in operator network A would need to be allowed to fetch an authentication vector from the subscriber's home operator B

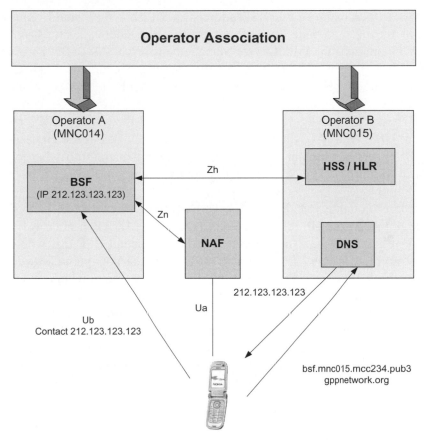

Figure 5.7. BSF node sharing in large networks. Reproduced by permission of © Nokia

subscriber database (i.e., HSS or HLR). Operators A and B are under the same administrative authority even if in different PLMNs. For this, the HSS / HLR would need to be configured to serve such an interoperator request coming from an external BSF. The actual messages and protocols can be the same as usually used within the network of one operator, i.e., the operator internal interface would become an external interface, but one that is still within one administrative domain.

Logical BSFs: The second problem is that due to the specific topology of a large network, an operator may have more than one logical BSF in their network. For example, a large operator may have a BSF in each province or region. So the address of the BSF in the Region 1 may be bsf.region1.mnc015.mcc234.pub.3gppnetwork. org. The current specifications of GAA, do not define how large networks should discover the correct BSF address if there are more than one. One approach to this problem is to preconfigure the BSF address in UE, but this would require extra support in the network side, e.g., provision the BSF address using Over-the-Air (OTA) mechanism or an OMA Device Management Server [OMADM] provisioning the BSF address. When UE starts GAA bootstrapping procedure, it will check if there is a preconfigured BSF address and use it if it is available.

Another solution is to have a single BSF portal or BSF Proxy (e.g., with the name bsf.mnc015.mcc234.pub.3gppnetwork.org) and it will forward the UE's request to the BSF in UE's HPLMN. The UE will use the HPLMN BSF received in bootstrapping as BSF_servers_domain_name. The problem is that the traffic burden of this BSF portal might be very heavy, especially during peak times, since it has to serve the whole network.

Yet another solution is to enable the BSFs to do redirections as part of the Ub reference point. As the Ub reference point is based on HTTP, a main BSF can use the HTTP redirect mechanism to redirect the UE to actual BSF that would handle the bootstrapping. In this case, the main BSF would be configured with logic that by taking a look at the received IMPI (or IMSI), will decide to which 'worker' BSF that particular UE should be redirected to. This has the same advantage as in the BSF proxy case above that the UE does not need to have any logic to decide which BSF to contact. Additionally, this approach does not have the traffic burden of the BSF proxy approach, where all Ub traffic is routed through this single proxy BSF. The downside to this approach is that not all UE implementations may support HTTP redirects.

References

[Apache]   Jakarta Apache Web Server. Available at http://httpd.apache.org/
[Bandit]   Bandit Project, (2007). Available at http://www.bandit-project.org/
[CMCC]   China Mobile Limited. Available at http://www.chinamobileltd.com/
[Heath06]   Craig Heath, *Symbian OS Platform Security*, Symbian Press, 2006.
[Higgins]   Higgins Project (2007). Available at http://www.eclipse.org/higgins/
[J2EE]   Java 2 Enterprise Edition (J2EE). Available at http://java.sun.com/j2ee
[Liberty]   Liberty Alliance Project. Available at http://www.projectliberty.org/
[LibertyIDFF]   Liberty Alliance Project, *Identity Federation Framework (ID-FF)*, Version 1.2. Available at http://www.projectliberty.org/resource_center/specifications/liberty_alliance_id_ff_1_2_specifications
[LibertyIDWSF] Liberty Alliance Project, *Identity Web Service Framework (ID-WSF)*, Version 2.0. Available at http://www.projectliberty.org/resource_center/specifications/liberty_alliance_id_wsf_2_0_specifications_including_errata_v1_0_updates
[OASIS-SAML]   Organization for the Advancement of Structured Information Standards (OASIS), *Security Asserion Markup Language (SAML)*, Version 2.0, March 2005. Available at http://www.oasis-open.org/specs/index.php#samlv2.0
[OASIS-WSSX]   Organization for the Advancement of Structured Information Standards (OASIS), *Web Service Secure Exchange (WS-SX) Technical Committee*. Available at http://www.oasis-open.org/committees/ws-sx

[OMABCAST]   Open Mobile Alliance (OMA), *OMA Mobile Broadcast Services*, Version 1.0 (2007). Available at http://www.openmobilealliance.org/

[OMADM]   Open Mobile Alliance (OMA), *OMA Mobile Device Management*, Version 1.2 (2007). Available at http://www.openmobilealliance.org/

[OpenID]   OpenID. Available at http://openid.net/

[RFC4279]   Internet Engineering Task Force (IETF), *Pre-Shared Key Ciphersuites for Transport Layer Security (TLS)*, RFC 4279, December 2005. Available at http://www.ietf.org/rfc/rfc4279.txt

[TR33.980]   3rd Generation Partnership Project (3GPP), Technical Report TR 33.980, *Liberty Alliance and 3GPP security interworking; Interworking of Liberty Alliance Identity Federation Framework (ID-FF), Identity Web Services Framework (ID-WSF) and Generic Authentication Architecture (GAA)*, Version 7.4.0 (2007). Available at http://www.3gpp.org/

[TS23.003]   3rd Generation Partnership Project (3GPP), Technical Specification TS 23.003, *Numbering, Addressing and Identification*, Version 7.4.0 (2007). Available at http://www.3gpp.org/

[TS24.109]   Generation Partnership Project (3GPP), Technical Specification TS 24.109, *Bootstrapping interface (Ub) and Network Application Function Interface (Ua); Protocol Details*, Version 7.5.0 (2006). Available at http://www.3gpp.org/

[TS29.109]   3rd Generation Partnership Project (3GPP), Technical Specification TS 29.109, *Generic Authentication Architecture (GAA); Zh and Zn Interfaces Based on the Diameter Protocol; Stage 3*, Version 7.5.0 (2006). Available at http://www.3gpp.org/

[TS33.220]   3rd Generation Partnership Project (3GPP), Technical Specification TS 33.220, *Generic Authentication Architecture (GAA); Generic Bootstrapping Architecture*, Version 7.7.0, (2007). Available at http://www.3gpp.org/

[TS33.221]   3rd Generation Partnership Project (3GPP), Technical Specification TS 33.221, *Generic Authentication Architecture (GAA); Support for Subscriber Certificates*, Version. 6.3.0 (2006). Available at http://www.3gpp.org/

[TS33.222]   3rd Generation Partnership Project (3GPP), Technical Specification TS 33.222, *Generic Authentication Architecture (GAA); Access to Network Application Functions Using Hypertext Transfer Protocol over Transport Layer Security (HTTPS)*, Version 7.2.0 (2006). Available at http://www.3gpp.org/

[TS33.246]   3rd Generation Partnership Project (3GPP), Technical Specification TS 33.246, *3G Security; Security of Multimedia Broadcast / Multicast Services*, Version 7.3.0 (2007).

6

Future Trends

In the first part of this chapter, we give a snapshot of the ongoing GAA work in 3GPP for Release 8 and beyond Release 8. In the second part, we conclude the book with a speculative view about potential outlook for GAA technology.

6.1 Standardization Outlook

Work on Release 8 of 3GPP is in progress at the time of writing of this book. Therefore, the concepts presented in this section may change substantially before the finalization of Release 8. We expect that in the final version of Release 8, GAA will be present with a larger range of specifications and applications than what is outlined here.

6.1.1 GBA Push

The new use cases of GAA in 3GPP and the adoption of GAA by other standardization bodies, like OMA, resulted in new service requirements for GAA. In particular, it became clear that, for some services, we can not make the assumption that the UE has always the possibility to connect to the BSF. Also, the MNO may have other reasons, e.g., load balancing on BSF, to push the security association information from the NAF to the UE. Therefore, 3GPP started to specify, so called, GBA Push function around the time when 3GPP Release 7 specification work was being finished. The GBA Push specification continued in Release 8 and will become a Release 8 feature.

Cellular Authentication for Mobile and Internet Services
Silke Holtmanns, Valtteri Niemi, Philip Ginzboorg, Pekka Laitinen and N. Asokan
© 2008 John Wiley & Sons, Ltd

The GBA Push functionality introduces a mechanism to bootstrap the security between a NAF and a UE, without forcing the UE to contact the BSF to initiate the bootstrapping. The GBA Push functionality builds on the architecture and functionality provided by [TS33.220]. Since the UE has not contacted the BSF to trigger the bootstrapping, it is necessary to define two new interfaces: one between the BSF and the NAF, and another one between the NAF and the UE. These new interfaces are mostly the same as defined in [TS33.220], but due to the changed baseline assumption, there are also some major differences. This was work in progress during the writing of the book and its latest status can be found in [TS33.223] document.

The GBA Push standardization in 3GPP was first triggered by the OMA when they expressed a need for a secure push mechanism where it should be possible to securely push data to the user using GAA. (It should be noted that e.g. WAP Push utilizes a nonsecure SMS message to start the process.) GBA Push can support the creation of a security association between a service in the network and a terminal in a broadcast network. 3GPP decided to study the usage scenarios provided by OMA so that such functionality can be used in a broad range of use cases. Below are use cases that were identified by OMA:

- Network-initiated service key refreshing and key distribution

    In normal GAA [TS33.220], the terminal starts the key agreement process. The network may wish to update the existing keys, e.g., for broadcasted content. The key creation material needed to render the protected content could be pushed to the terminal at regular intervals to avoid having long-lived keys in the terminal. MBMS Security [TS33.246] has its own inbuilt key refresh mechanism. But other services may want to create a new security association with the terminal as well, and for example, for performance reasons, MBMS may benefit from having an alternative method available. For instance, it is important to guarantee that fresh keys are available when the old ones expire and to avoid service consumption interruptions when fresh keys need to be requested by the terminal. A push key distribution mechanism also enables load balancing in the service network as the user can receive a new security association well in advance before the peak load times. Also, some security-sensitive services may want to be absolutely sure that the used keys are indeed fresh.

- Distribution of tokens or tickets

    The idea behind this use case is that the terminal obtains a token that needs to be presented to a service provider over a connection with a return channel, e.g., IP connection. Hence, in this case, some messages indicating successful key delivery or error are possible. A particular kind of token is a ticket, e.g., for a concert or another type of event. The delivery of such a ticket needs to be secured to avoid copying. If the security association is established well before the actual event, then the actual ticket distribution could be scheduled in a predefined time window. This may help to avoid problems, like accidental deletion, copying, etc. For low-value tokens or tickets, it may also be interesting to enable delivery when the terminal has no IP connectivity. Also, if the actual distribution channel is reliable enough for the use case, then the back channel usage to report the successful delivery is not needed.

- Network-initiated services

    Normal GAA starts with the user contacting a server, but there are quite a number of services that are actually triggered from the network. Typically the server 'triggers' the terminal to contact the server. Often, SMS is used for this, but SMS is not a secure trigger. Due to the lack of other mechanisms, some OMA enablers, like SUPL, DM, and Download / DRM are using this kind of approach. OMA BCAST group is already using normal GAA in their Smartcard Profile for the service protection. OMA BCAST Smartcard Profile could use network-initiated registration and delivery of long-term keys enabled by a GBA Push. GBA Push could also potentially protect against replay and DoS attacks in some cases.

GBA Push is currently designed to address these use cases. After it its specification is finalised GBA Push will be one of GAA building blocks.

The major differences between normal GAA and GBA Push are the new interfaces. (See Figure 6.1.) The Upa interface between NAF and terminal is used to push the security association to the terminal and it replaces the Ub interface between the terminal and the BSF. The modified Zn interface, which now needs to transport slightly different information, e.g., the indication of user identity by NAF, is called Zpn if it is diameter-based or Zpn' if it is web service-based. The new interface builds, as much as possible, on the existing Zn interface.

6.1.2 GAA User Privacy

Since the basic GAA functionality is used as an enabler for more and more services, it was decided to add an additional privacy protection to the GAA functionality for Release 8. GAA has been recently taken up by other standardization bodies, but originally, GAA was designed with the already deployed 3GPP security environment in mind. For example, it was assumed that the underlying network bearer is encrypted, hence, the user's private identifier IMPI does not need additional encryption to protect the user from identity sniffing. But with the bearer-independent usage of GAA driven by different standardization organizations, there was an agreement made in 3GPP that an additional layer of protection for users' privacy is needed to

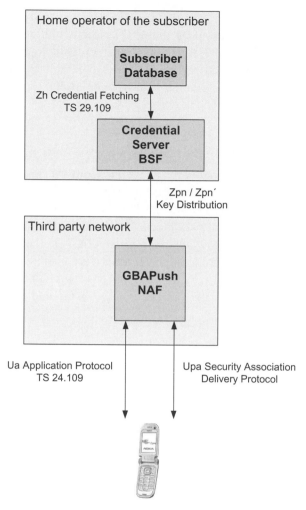

Figure 6.1.   GBA push architecture. Reproduced by permission of © Nokia

accommodate also usage of GAA's Ub interface over arbitrary IP-based networks
that may offer no encryption, e.g., WLAN hotspots. Without this additional user
privacy protection, an eavesdropper may trace a user and link together different user
activities.

To address this concern, a temporary identifier was introduced. It is called the
Temporary IP Multimedia Private Identity (TMPI) and it is for use on the Ub reference point. The idea is to use for each new bootstrapping run a new temporary user
identifier, hence, making linking impossible. The TMPI is created locally in the BSF

and the terminal for each bootstrapping run in order to be used for the next bootstrapping procedure with the network. The terminal would start the next Ub run with the TMPI and the BSF would recognize the structure of the identifier. If the terminal is not successful using the TMPI, it would then make another try with the IMPI. The terminal and the BSF need to indicate to each other that they support this feature by using product tokens in HTTP headers of Ub interface.

The TMPI mechanism provides protection against a passive eavesdropper, but not against an active attack, where the attacker forces the terminal to revert back to using IMPI instead of TMPI. The protection against an active attack would require substantial changes to the protocols and would also introduce serious compatibility issues to previous Releases of GAA. The protection level can be compared to the TMSI mechanism used in GSM and UMTS networks today. Also in that context, effective protection mechanisms against active attacks have been estimated to be too costly.

6.1.3 GAA in Evolved Packet Systems (EPSs) and Mobile IP (MIP)

Access independence of the GAA has the following advantage: GAA can work over any access network technology and works in similar way in all cases. This covers also access technologies that are not yet deployed or even fully defined. One notable example is the evolved UTRAN (E-UTRAN) specified in 3GPP Release 8. This new radio technology uses multiantenna technologies and new transmission schema that are different from the ones used in GSM, CDMA2000 or WCDMA. It is also referred to as LTE (Long Term Evolution), and it offers bit rates comparable to the current bit rates in fixed broadband connections.

Access independence of the GAA has also a disadvantage: IP communication has to be established before GAA authentication can be used. This is the reason why GAA could not be used for access authentication towards E-UTRAN.

Note, however, that in principle, GAA could be used for authenticating secondary radio access, under the assumption that GAA bootstrapping has already been done over primary radio access. For instance, primary radio access could be cellular access, whereas the secondary access would be over short-range radio, e.g., WLAN or Bluetooth.

Anyway, the security aspects of the evolution of 3G networks are currently being studied. The threats and potential countermeasures are identified. In the EPS that lays the path for the future networks also, GAA is considered for Mobile IP (MIP) security and for protection of user location. MIP itself is needed for making it possible to connect also non-3GPP-based access networks to the EPS.

The basic idea is that the terminal needs security credentials specifically for MIP signalling security. But to optimize the system, the needed security association

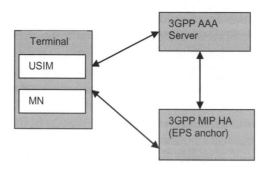

Figure 6.2.   GAA and mobile IP

should be derived from a security credential already available, i.e., the cryptographic keys stored on the USIM. The terminal acts as a Mobile Node (MN) in MIP context.

The MN needs to authenticate towards the networks and vice versa. A subscriber that uses Mobile IP will have personal data in a subscriber profile stored in the subscriber database (AAA server) of the operator. For privacy and security reasons, this data should only be disclosed to the authorized parties. The subscriber profile will be used at MIP registration and change of location between different access networks to determine if the subscriber is authorized to continue. The [TR33.922] outlines three GAA-based approaches how to secure MIP. Note, however, that so far, none of these approaches have been chosen for further specification work related to aspects of non-3GPP access technologies in EPS.

In MIP there exist three security associations, as outlined in Figure 6.2.

The basis is the trust relationship between the terminal and the 3GPP AAA server that is based on the user subscription. Hence, we assume that the AAA server in the user's home network (Home – PLMN) is responsible for the user authentication and authorization, for example, using the AKA Protocol. The MIP authentication is independent from the access authentication to the actual network and can be seen like a service authentication with some particularities. Therefore, GAA can be used for MIP key provisioning.

The second trust relationship is between the 3GPP Home Agent (HA) and the terminal. This is needed so that the HA can perform necessary operations for the terminal's mobility requirements. This trust relationship between the terminal and the HA is usually established dynamically without preprovisioned shared secret.

The third trust relationship is between two network nodes, the MIP HA and the AAA server. In the non-roaming case these two network nodes reside in the same operator network. In the roaming case, the AAA proxy would be used in the visited network and the AAA server in the home network. In the roaming case, the involved network operators have a contractual agreement, hence, this trust relationship can be assumed for both cases. Interoperator communication is often secured with Network Domain Security [TS33.210] or operator-specific means, e.g., based on IPsec.

If a Mobile IPv4 Foreign Agent (FA) is used, then Figure 6.2 needs to be extended to accommodate the roaming nodes and [RFC3344] can be used for the interface security of those interfaces between the roaming nodes. The key that secures the association between the terminal, in particular, the MN, and the HA, can utilize the key generation and key distribution as outlined in [RFC3957]. If this approach is used, then a pre-shared secret between the terminal and the AAA server is needed to establish another shared secret between the terminal (MN) and the HA, or for the roaming case, between the MN and the FA. GAA could be used to establish this shared secret.

6.2 Outlook for GAA

Having discussed the technical details, variations and possible applications of GAA, let us take a step back and speculate on the potential outlook for GAA as a technology. The prevalent authentication technology for Internet services today is the use of username and passwords. In order to be successful, any challenger technology must offer tangible advantages over the incumbent. Therefore, to analyze the prospects of GAA, we first begin with the strengths and weaknesses of the incumbent technology. The biggest strengths of username / password authentication are:

- **Familiarity**: Anyone who is an Internet user today already has, or can quickly gain, a mental model of what a password is. A large number of users can start using usernames / passwords without any training.
- **Deployability**: When Internet services enrol users into their password-based authentication systems, a majority of them trade-off security in return for easier deployability. Typically, users are allowed to create an account via a web front-end and claim an e-mail address. The activation of the new account is then subject to an e-mail routability check to the claimed address. The server-side software is readily available as part of most web site building tools. No new infrastructure or client-side software is needed.
- **Portability**: As long as the number of passwords is small, users carry their passwords in their head and can thus use them from any client device: be it their own home or office computers, other portable devices, or other computers that they may be using temporarily.

The primary weaknesses of authenticating with username / password are:

- **Usability**: The usability drawbacks of passwords are well known: users have difficulties in choosing strong passwords, remembering them and entering them correctly when prompted.

- **Security**: The usability difficulties as well as the trade-offs made in order to ease deployment lead to reduced levels of security. Users pick weak, easy-to-guess passwords. E-mail routability checks, typically part of the enrolment process of many popular web services, can be intercepted easily by anyone along the path of the e-mail and accounts can be hijacked. But the most serious security issue with the username / password approach is its vulnerability to phishing. Attackers can easily fool users into revealing their passwords to the wrong server. Phishing is an enormous problem for Internet services today.

As we have seen already, GAA can significantly ease the usability burden. GAA-based authentication lends itself to an intuitive mental model of using a mobile device as a physical 'key' to gain access to services.

GAA can also effectively address the security shortcomings of username / passwords. A well-designed GAA-based authentication mechanism can protect users against phishing. For example, combining a GAA-based HTTP-digest authentication with server-authenticated TLS (Section 4.1.1) with appropriate protection against man-in-the-middle (as is done in Section 3.3.3) can thwart phishing and pharming attacks.

This leaves us with two aspects of GAA, which will determine whether it can effectively compete with the username / password approach: **deployability** and **portability**.

Deployability: Successful deployment and use of GAA require the following:

- device manufacturers releasing GAA-capable mobile devices;
- mobile operators setting up BSFs; and
- mobile operators making it easy for service providers to use NAFs.

Lead GAA applications, like broadcast mobile TV, are paving the way for the first two requirements to be met. In Nokia S60 devices, GAA functionality can be added via a downloadable software package. Setting up and operating a BSF may incur a significant initial cost. However, once a BSF is deployed for a particular network application, the incremental cost of adding new applications is low. The technical effort in setting up an NAF is made easier by the availability of easy-to-use libraries, like the Java NAF library referred to in Section 5.1.3, which already exist. Thus, we can expect that operators deploying a BSF for lead GAA applications will quickly follow that with additional applications for operator-provided services.

Third-party service providers wishing to use GAA must adapt their application to use GAA, and sign up with one or more operators for the right to use their BSFs. Availability of NAF libraries makes the task of adapting applications easy. To enable widespread adaptation of GAA by third-party service providers, it is important that operators make it easy for service providers to sign up with the operator to use its

BSF. Automated, web-based sign-up mechanisms would go a long way in encouraging service providers to use GAA.

Thus, the critical factors in successful deployment of GAA are whether and how many operators install BSFs and how easy they make it for service providers to start using their BSFs.

The relationship with subscribers is perhaps the biggest asset operators have. While any party could provide services to mobile subscribers, only the operators have the ability to authenticate the subscribers in a scalable and secure fashion. GAA constitutes a new way to monetize this asset. Therefore, we could expect that operators will make the investment needed to do so.

Portability: To be really successful, GAA-based authentication must be usable on devices like PCs and Internet tablets, which are not cellular devices. Although they could contain smart card readers and the users could achieve portability by moving their UICCs from device to device, this suffers from poor usability and is therefore an unlikely usage scenario. What is needed is an easy way for users to carry out GAA-based authentication from their noncellular devices by effectively exploiting the GAA capabilities on their cellular devices. This is keeping with the idea of a personal trusted device acting as a key. The split-terminal model of GAA discussed in Section 3.5.5 is intended for this purpose. There are different ways of configuring the split-terminal approach. They differ in terms of security, usability and availability of standardized mechanisms.

Initial deployment of GAA seems to be starting. In the preceding chapters, we have tried to explain the great potential of GAA in addressing one of the vexing problems on the Internet: easy and effective user authentication. However, it is too early to tell how much of this potential will be realized.

At this point, as we reach the end of this book, the astute reader would have noticed that even though the core idea behind GAA is very simple, it took a book to describe its various intricacies. This is neither surprising nor unique to GAA. To paraphrase Thomas Edison, any successful deployment of technology comes out of 1% inspiration and 99% perspiration. What might seem like a simple and elegant concept at first glance would take a number of practical details to be worked out before it can be widely deployed. In the process, the intended applications and the concept itself may change shape. A good example is public key technology. When it was first conceived in the 1970s, it appeared to be an ingenious, aesthetically pleasing concept, that would be relatively straightforward to deploy. Thirty years later, the difficulties of deploying public key technology as originally intended have become apparent. Widespread deployments of public key technology, as in the use of server-authenticated web browsing with TLS / SSL, bear little resemblance to the originally envisaged usage scenarios. In the course of being adapted to address the deployability and portability challenges of GAA, it, too, will change shape and be used in ways we have not imagined so far.

References

[MSF]   MultiService Forum (MSF). Available at http://www.msforum.org/

[RFC3344]   Internet Engineering Task Force (IETF), *IP Mobility Support for IP v4,* RFC 3344, August 2002. Available at http://www.ietf.org/rfc/rfc3344.txt

[RFC3957]   Internet Engineering Task Force (IETF), *Authentication, Authorization, and Accounting (AAA) – Registration Keys for Mobile IPv4,* RFC 3957, March 2005. Available at http://www.ietf.org/rfc/rfc3957.txt

[RFC4004]   Internet Engineering Task Force (IETF), *Diameter Mobile IP v4 Application,* RFC 4004, August 2005. Available at http://www.ietf.org/rfc/rfc4004.txt

[TR33.821]   3rd Generation Partnership Project (3GPP), Technical Report TR 33.821, *Rationale and Track of Security Decisions in Long Term Evolved (LTE) RAN / 3GPP System Architecture Evolution (SAE),* Version 0.3.0 (2007). Available at http://www.3gpp.org/

[TR33.922]   3rd Generation Partnership Project (3GPP), Technical Report TR 33.922, *Security Aspects for Inter-Access Mobility between non 3GPP and 3GPP Access Network,* Version 0.0.3 (2007). Available at http://www.3gpp.org/

[TS24.109]   3rd Generation Partnership Project (3GPP), Technical Specification TS 24.109, *Bootstrapping interface (Ub) and network application function interface (Ua); Protocol details,* version 7.5.0 (2006), http://www.3gpp.org/

[TS29.109]   3rd Generation Partnership Project (3GPP), Technical Specification TS 29.109, *Generic Authentication Architecture (GAA); Zh and Zn Interfaces based on the Diameter protocol; Stage 3,* Version 7.5.0 (2006). Available at http://www.3gpp.org/

[TS33.210]   3rd Generation Partnership Project (3GPP), Technical Specification TS 33.210, *3G security; Network Domain Security (NDS); IP Network Layer Security,* Version 7.2.0 (2006). Available at http://www.3gpp.org/

[TS33.220]   3rd Generation Partnership Project (3GPP), Technical Specification TS 33.220, *Generic Authentication Architecture (GAA); Generic Bootstrapping Architecture,* Version 7.7.0, (2007). Available at http://www.3gpp.org/

[TS33.223]   3rd Generation Partnership Project (3GPP), Technical Specification TS 33.223, *Generic Authentication Architecture (GAA); Generic Bootstrapping Architecture (GBA) Push Function,* Version 0.5.0 (2007), Release 8.

[TS33.246]   3rd Generation Partnership Project (3GPP), Technical Specification TS 33.246, *3G Security; Security of Multimedia Broadcast / Multicast Services,* Version 7.3.0 (2007)

Terminology and Abbreviations

For official terminology and abbreviations of 3GPP, see 3GPP TS21.905.

Terminology

3GPP   3rd Generation Partnership Project: An organization, established in 1998 that works on 3rd generation mobile standardization. The partners are telecommunication standards bodies from Asia, America and Europe. At the time 3GPP was established, it was agreed to use WCDMA technology for the radio network and the GSM architecture for the core network.

3GPP2   3rd Generation Partnership Project 2: An organization that performs 3rd generation mobile standardization work from the IS-95 radio technology basis.

AKA   Authentication and Key Agreement Protocol: AKA has been designed for securing subscriber access to UMTS [TS33.102]. The authentication part of the AKA protocol is needed to verify the subscriber's identity while key agreement is used for generating keys CK and IK that are subsequently used in encryption of traffic in the radio network (CK) and also for protecting integrity of the signalling messages (IK). A notable difference between AKA and the GSM authentication is that AKA provides *mutual* authentication: In UMTS, terminal and network authenticate each other, while in GSM, only the network authenticates the terminal.

Cellular Authentication for Mobile and Internet Services
Silke Holtmanns, Valtteri Niemi, Philip Ginzboorg, Pekka Laitinen and N. Asokan
© 2008 John Wiley & Sons, Ltd

AuC   Authentication Center: AuC is the network element responsible for storing the permanent cryptographic keys bound to the subscription and computing the session keys and the authentication data. The AuC is usually implemented as an integral part of HLR or, in later releases of the 3GPP specifications, HSS.

BSD/A   BCAST Service Distribution / Adaptation: A broadcast mobile TV server that, among other things, encrypts TEK with SEK/PEK.

BSF   Bootstrapping Server Function: BSF is a new network function introduced in GAA. It facilitates the use of AKA to bootstrap a new GAA master session key Ks.

BSM   BCAST Subscription Management: A broadcast mobile TV server that, among other things, delivers SEK/PEK to the mobile device in response to subscriber's request for a program or service. Since that message exchange is secured with GAA, BSM includes a NAF component.

CDMA   Code Division Multiple Access: CDMA is a spread spectrum radio technology that allows multiple users to be multiplexed over the same physical channel. The cellular network standards IS-95 and CDMA2000 are often referred to as 'CDMA'.

CDMA2000   3GPP2 standard that defines the 3G evolution of IS-95.

CK   Ciphering Key. See AKA.

E-UTRAN   Evolved UTRAN: New radio technology defined in 3GPP. E-UTRAN uses multiantenna technologies and new transmission schema that are different from the ones used in GSM, CDMA2000 or WCDMA. It is also referred to as LTE (Long-Term Evolution) and it offers bit rates comparable to the current bit rates in fixed broadband connections.

GAA   Generic Authentication Architecture. GAA consists of a set of specifications that describe how the cellular security infrastructure can be used to provide a general-purpose authentication service for applications and services. It has been standardized both in the 3rd Generation Partnership Project (3GPP) and its North American counterpart, the 3rd Generation Partnership Project 2 (3GPP2). GAA includes GBA, Support for Subscriber Certificates (SSC), Access to application servers with HTTPS and Key Centre

GBA   Generic Bootstrapping Architecture: a fundamental building block of GAA. GBA enables automatic provisioning of shared keys between the mobile terminal and an application server (provided that the user has a valid subscription to cellular network services). Other GAA constituents are built on top of GBA.

GSID   GAA Service Identifier: An identifier of a User Security Settings (USS) entry.

GUSS	GBA User Security Settings. The GAA-specific data of a subscriber that is stored in HSS. User Security Settings (USS) containing identities and authorizations of the subscriber for a particular service may be included in GUSS.
HLR	Home Location Register: HLR is a subscriber database in GSM architecture that contains the subscriber profiles and information about the subscriber's physical location. In addition, HLR obtains authentication data of subscribers from an Authentication Centre (AuC) and forwards the data to other network elements.
HS	Home Server: A generic term for servers like HLR and HSS.
HSS	Home Subscriber Server: HSS is a subscriber database in UMTS architecture that is similar to the GSM Home Location Register (HLR). Often the HSS is implemented as an HLR with extended functionality.
IK	Integrity Key. See AKA.
IMPI	IP Multimedia Private Identity: IMPI is a Uniform Resource Identifier (URI) that can consist of digits or alphanumeric identifiers (e.g., subcriber@operatorID.com).
IMSI	International Mobile Subscriber Identity: A unique number associated with UMTS subscriber. IMSI consists of MCC (Mobile Country Code), followed by MNC (Mobile Network Code) and MSIN (Mobile Station Identification Number) and is typically 15-digit long.
IS-95	2G CDMA standard, also called cdmaOne.
LTE	Long-Term Evolution: see E-UTRAN.
LTKM	Long-Term Key Message: In broadcast mobile TV, LTKM is the message that is delivered point-to-point from BSM to the mobile terminal and containing service or program key SEK/PEK.
KDF	Key Derivation Function. A function that derives cryptographic keys from its inputs. The inputs may include, e.g., other cryptographic keys and text strings. The KDF used in GAA consists of an application of HMAC-SHA-256 and its exact form is defined in Annex B of [TS33.220].
MBMS	Multimedia Broadcast Multicast Service. The 3GPP MBMS specification defines enhancements to the GPRS bearer to provide multicast and broadcast capability. In essence, MBMS defines how to multicast and broadcast data over 3G cellular radio network.
MCC	Mobile Country Code: The first three digits of an IMSI number that identify subscriber's country.
ME	Mobile Equipment. See UE.
MGV-S/F	MBMS key Generation and Validation Storage and Function: In MBMS and broadcast mobile TV, MGV-S/F is application on the

	UICC that stores long-term keys (SEK/PEK) and decrypts the short-term keys (TEK).
MNC	Mobile Network Code: Part of the IMSI that identifies subscriber's home network. MNC follows the MCC and consists of either two or three digits.
MSIN	Mobile Station Identity Number: Part of the IMSI that identifies the mobile station in the home network customer database (9–10 digits).
MSISDN	Mobile Subscriber International ISDN Number: MSIDN is the telephone number of the mobile subscriber, while IMSI is the key to MNO subscribers database.
NAF	Network Application Function: NAF is an application server that is able to use shared keys produced by GAA to authenticate subscribers. The NAF functionality is typically only a small part of the server software and can be, for example, realized through a NAF library.
PEK	Program Encryption Key. See LTKM.
RNC	Radio Network Controller: RNC is the switching and controlling element of UTRAN.
SEK	Service Encryption Key. See LTKM.
STKM	Short-Term Key Message: In broadcast mobile TV, STKM is the message that is broadcasted together with the content and containing short-lived traffic encryption key TEK. The broadcasted TEK is itself encrypted with service or program key SEK/PEK.
TMSI	Temporary Mobile Subscriber Identity: To improve identity privacy TMSI is generated per geographical location; the mobile device can use TMSI during signalling instead of its permanent identifier, IMSI.
UE	User Equipment: In 3G network, UE is the official name of user's terminal. UE is often called mobile station or mobile device. It consists of the end-user independent part called Mobile Equipment (ME), the Terminal Equipment (TE) part that provides end-user application functions, and USIM. USIM is the user-dependent part of the mobile terminal that is implemented as an application on UICC: a smart card that is physically inserted into the mobile terminal.
UICC	Universal Integrated Circuit Card: UICC may contain one or more USIMs and other applications, like ISIM and MGV-S/F; see also UE.
UMTS	Universal Mobile Telecommunication System. UMTS is the leading 3rd generation mobile communication system.
USS	User Security Settings: Part of GUSS that contains identity and authorization of the subscriber for a particular service. (e.g., the authorization flag in the USS may permit or disable the particular service for the subscriber.)

UTRAN   Universal Terrestrial Radio Access Network. UTRAN refers to UMTS with WCDMA radio access network.

Abbreviations

3GPP	3rd Generation Partnership Project
3GPP2	3rd Generation Partnership Project 2
AAA	Authentication, Authorization and Accounting
ACL	Access Control List
AKA	(UMTS) Authentication and Key Agreement Protocol
AP	Authentication Proxy
ARIB	Association of Radio Industries and Businesses
ATIS	Alliance for Telecommunications Industry Solutions
AuC	Authentication Centre
AV	(AKA) Authentication Vector
BCAST	(OMA) Broadcasting Working Group
BS	Base Station
BSD/A	BCAST Service Distribution / Adaptation
BSF	Bootstrapping Server Function
BSM	BCAST Subscription Management
CA	Certificate/Certification Authority
CAVE	Cellular Authentication and Voice Encryption
CCSA	Canadian Cable Systems Alliance
CDMA	Code Division Multiple Access
CK	Ciphering Key
CS	Circuit Switched
DM	Device Management
DNS	Domain Name Server
DoS	Denial of Service
DRM	Digital Rights Management
DVB-H	Digital Video Broadcasting – Handheld
EAP	Extensible Authentication Protocol
EPS	Evolved Packet System
ESN	Electronic Serial Number
ETSI	European Telecommunications Standards Institute
EV-DO	Evolution – Data Only or Data Optimized
FA	Foreign Agent
FQDN	Fully Qualified Domain Name
GAA	Generic Authentication Architecture
GBA	Generic Bootstrapping Architecture
GEA	GPRS Encryption Algorithm

GSID   GAA Service Identifier
GSM   Global System for Mobile communication
GUSS   GBA User Security Settings
HA   Home Agent
HLR   Home Location Register
HS   Home Server
HSS   Home Subscriber Server
HTTP   Hypertext Transfer Protocol
HTTPS   Hypertext Transfer Protocol over Transport Layer Security
ICCID   Integrated Circuit Card ID
I-CSCF   Interrogating Call Session Control Function
ID-FF   Identity Federation Framework
IdP   Identity Provider
IETF   Internet Engineering Task Force
IK   Integrity Key
IKE   Internet Key Exchange protocol
IMPI   IP Multimedia Private Identity
IMS   IP Multimedia Subsystem
IMSI   International Mobile Subscriber Identity
IP   Internet Protocol
IPsec   Internet Protocol security
ISDN   Integrated Services Digital Network
ISIM   IP Multimedia Subsystem Identity Module
ITU   International Telecommunication Union
J2EE   Java 2 Enterprise Edition
LTE   Long-Term Evolution
LTKM   Long-Term Key Message
KDC   Key Distribution Center
KDF   Key Derivation Function
M3UA   Message Transfer Part 3 (MTP3) – User Adaptation Layer
MAC   Message Authentication Code
MAP   Mobile Application Part
MBMS   Multimedia Broadcast Multicast Service
MCC   Mobile Country Code
ME   Mobile Equipment
MGV-S/F   MBMS key Generation and Validation Storage and Function
MIKEY   Multimedia Internet KEYing
MIP   Mobile IP
MN   Mobile Node
MNC   Mobile Network Code
MNO   Mobile Network Operator
MPEG   Moving Picture Experts Group

MRK	MBMS Request Key
MSF	MultiService Forum
MSIN	Mobile Station Identity Number
MSISDN	Mobile Subscriber International ISDN Number
MTP	Message Transfer Part
MUK	MBMS User Key
NAF	Network Application Function
NAI	Network Access ID
NASS	Network Attachment Subsystem
NAT	Network Address Translation
NDS	Network Domain Security
OASIS	Organization for the Advancement of Structured Information Standards
OMA	Open Mobile Alliance
OSI	Open System Interconnection
PAG	(OMA) Presence and Availability working Group
PEK	Program Encryption Key
PIN	Personal Identification Number
PKI	Public Key Infrastructure
PLMN	Public Land Mobile Network
PS	Packet Switched
PSK TLS	Pre-Shared Key Transport Layer Security
PW	Password
QOP	Quality Of Protection
RA	Registration Authority
RAN	Radio Access Network
RADIUS	Remote Authentication Dial-In User Service
RNC	Radio Network Controller
RP	Relying Party
RTP	Real Time Transport Protocol
R-UIM	Removable User Identity Module
SAML	Security Assertion Markup Language
SAP	SIM Access Profile
SCCP	Signalling Connection Control Part
S-CSCF	Serving Call Session Control Function
SCTP	Stream Control Transmission Protocol
SEK	Service Encryption Key
SELinux	Security Enhanced Linux
SGSN	Serving GPRS Support Node
SID	Secure Identifier
SIM	Subscriber Identity Module
SIP	Session Initiation Protocol

SLF	Subscriber Location Function
SMS	Short Message Service
SOAP	Simple Object Access Protocol
SP	Service Provider
SQN	(AKA) Sequence Number
SSC	(GAA) Support for Subscriber Certificates
SSO	Single Sign On
STS	Security Token Service
STKM	Short Term Key Message
SUPL	Secure User Plane Location
TCAP	Transaction Capabilities Application Part
TCG	Trusted Computing Group
TCP	Transport Control Protocol
TEK	Traffic Encryption Key
TISPAN	Telecommunications and Internet converged Services and Protocols for Advanced Networking
TLS	Transport Layer Security
TMPI	Temporary IP Multimedia Private Identity
TMSI	Temporary Mobile Subscriber Identity
TTA	Telecommunication Technology Association
TTC	Telecommunication Technology Committee
UDP	User Datagram Protocol
UE	User Equipment
UICC	Universal Integrated Circuit Card
UIM	User Identity Module
UMTS	Universal Mobile Telecommunication System
USIM	Universal Subscriber Identity Module / UMTS Subscriber Identity Module
USS	User Security Settings
VID	Vendor Identifier
VLR	Visited Location Register
VPN	Virtual Private Network
W3C	World Wide Web Consortium
WAP	Wireless Application Protocol
WIM	Wireless Identity Module
WLAN	Wireless Local Area Network
WPKI	Wireless PKI
WSDL	Web Service Description Language
WSF	Web Service Framework
WS-SX	Web Service Exchange Service
XCAP	XML Configuration Access Protocol
XDM	XML Document Management

Index

Cellular Authentication for Mobile and Internet Services
Silke Holtmanns, Valtteri Niemi, Philip Ginzboorg, Pekka Laitinen and N. Asokan
© 2008 John Wiley & Sons, Ltd